Berry Crops Production: Cultivation, Breeding and Health Benefits

Berry Crops Production: Cultivation, Breeding and Health Benefits

Guest Editor

Toktam Taghavi

Basel • Beijing • Wuhan • Barcelona • Belgrade • Novi Sad • Cluj • Manchester

Guest Editor
Toktam Taghavi
Agricultural Research Station
Virginia State University
Petersburg
USA

Editorial Office
MDPI AG
Grosspeteranlage 5
4052 Basel, Switzerland

This is a reprint of the Special Issue, published open access by the journal *Horticulturae* (ISSN 2311-7524), freely accessible at: https://www.mdpi.com/journal/horticulturae/special_issues/berry_crops_production.

For citation purposes, cite each article independently as indicated on the article page online and as indicated below:

Lastname, A.A.; Lastname, B.B. Article Title. *Journal Name* **Year**, *Volume Number*, Page Range.

ISBN 978-3-7258-3637-6 (Hbk)
ISBN 978-3-7258-3638-3 (PDF)
https://doi.org/10.3390/books978-3-7258-3638-3

Contents

About the Editor

Toktam Taghavi

Toktam Taghavi holds a Ph.D. in Horticultural Science from the University of Tehran, where her research focused on plant nutrition and the postharvest physiology of strawberries. She has established herself as a leader in the field through extensive experience in plant production, postharvest research, and the molecular genetics of horticultural crops. Currently, she serves as an Associate Professor of Horticulture at Virginia State University, where she directs the Postharvest Research Program. Her work aims to develop sustainable, eco-friendly postharvest practices to extend the shelf life of fresh fruits and vegetables, implementing cutting-edge research to enhance food quality and reduce waste.

 *horticulturae*

Editorial

Berry Crops Production: Cultivation, Breeding and Health Benefits

Toktam Taghavi

Agricultural Research Station, Virginia State University, Petersburg, VA 23806, USA; ttaghavi@vsu.edu

1. Introduction

Horticulture is among the most intensive agricultural systems [1]. The horticulture industry is undergoing rapid changes due to environmental, economic, and social constraints, and the berry industry is no exception [2]. Climate change, migration, economic instabilities, and public health concerns [3] have been significant challenges in recent years. The science is also changing rapidly, as new technologies such as genome sequencing, bioinformatics, automation, and sensing and imaging technologies have emerged [4]. These technologies have led to the creation of large amounts of data in different areas of research; therefore, the results must be commercialized and incorporated into the established breeding, production, handling, and storage practices. This Special Issue brings together information on genetic and environmental conditions affecting small fruit production and quality. New technologies that could be adapted to help us improve production quality have also been explored.

2. An Overview of Published Articles

This Special Issue captures the diversity of the studies focusing on genetic and environmental factors affecting small fruit production and quality and the new technologies that can be adopted. It contains ten articles, which I will describe briefly in the following paragraphs. It is also worth clarifying that the purpose of this editorial is not to elaborate upon each of the texts, but rather to encourage the reader to explore them.

Contribution 1 demonstrates that the soil incorporation of sunn hemp three weeks before strawberry planting did not benefit strawberry plant growth or fruit yield, and lowered the level of total soil N, leading to lower above-ground plant dry weight and a lower accumulation of N, P, and K by the end of strawberry season. However, no significant differences in the whole-season marketable fruit yield were observed due to the sunn hemp treatment compared to the control nutrient management treatment. This study confirmed the importance of long-term studies to understand the soil nutrient dynamics and plant nutrient uptake as affected by sunn hemp incorporation, which is also applicable to other green cover crops incorporated into the soil.

The second article, by Patel et al. (contribution 2), is focused on variations in fruit quality and anthocyanin content in ten strawberry cultivars grown in high tunnels and open fields during the harvest season. They also used an automatic titrator and a digital pocket acidity meter to compare the titratable acidity data. Genetic differences among cultivars were the primary variable and significantly affected total soluble solids (TSS), titratable acidity (TA), acidity, and total anthocyanin content. There was a weak correlation between TA and acidity, measured by the pocket acidity meter. The results emphasize the effect of genotype on fruit quality. The article also suggests using a pocket acidity meter for open-field strawberry production if a fast and less expensive method for acid assessment is desired.

The article by Marcellini et al. (contribution 3) studied the effect of three irrigation regimes (standard, 20% less, and 40% less) on three remontant strawberry cultivars ('Albion', 'San Andreas', and 'Monterey'). 'Monterey' had the best fruit quality, folate, and

Citation: Taghavi, T. Berry Crops Production: Cultivation, Breeding and Health Benefits. *Horticulturae* 2024, 10, 875. https://doi.org/10.3390/horticulturae10080875

Received: 21 June 2024
Revised: 12 August 2024
Accepted: 15 August 2024
Published: 19 August 2024

antioxidant content at 20% less irrigation without significant yield reduction. Reducing the water by 20% did not significantly affect the yield, fruit quality, or nutrients. The data suggested that less than standard irrigation and more water management can save water by 226 m^3 per hectare per cultivation cycle.

The fourth article, by Čabilovski et al. (contribution 4), compared organic vs. mineral fertilization of strawberries. The highest yield was obtained from mineral NPK fertilizers across all three years of the study. Applying organic fertilizers (composts) improved fruit quality (TSS, antioxidants, and anthocyanins) in the second and third years of production. The results emphasize the importance of the time needed for organic fertilizers to affect strawberry fruit quality and the lack of a possible effect on strawberry yield in a short term.

The article by Mbarushimana et al. (contribution 5) tested the effect of a high tunnel structure on eight strawberry cultivars compared to open-field conditions and the economic feasibility (gross and net revenues) of both approaches over three marketing strategies. The open field had higher net revenues for all marketing strategies due to the higher yield and lower production costs. The authors suggested that growers should focus on open-field rather than high-tunnel production in the commonwealth of Virginia, US.

The sixth article, by Hosainpour et al. (contribution 6), developed a machine vision system for the quality grading of white mulberry fruits using a digital camera. Using artificial neural networks and support vector machine classifiers, they identified 23 features to classify the fruits in two grades (high and low). The model's high accuracy confirms that it is a reliable, low-cost tool for monitoring the quality of dried white mulberries.

The article by Zydlik et al. (contribution 7) studied the effect of foliar fertilization with calcium and microelements on the yield and fruit quality of highbush blueberry. Yield, fruit firmness, and TSS increased. However, the foliar fertilization did not affect chlorophyll, fruit coloration, acids, and phenolic contents of highbush blueberries.

In the article by Rivero et al. (contribution 8), different temperatures and photoperiods were tested on the flowering and yield of everbearing strawberry cultivars. They discovered that long days at 15–21 °C enhanced flowering and short days at 27 °C decreased flowering. Runner formation was inversely related to flowering and was enhanced by short days. They concluded that the fruits are strong sinks for photosynthates, and significant fruiting flushes create long off periods and small second flushes. Therefore, the size of the first fruit flush must be compromised to optimize the fruit yield of the second flush or the season crop distribution.

The article by Zhang et al. (contribution 9) examined the influence of genotype and harvest date on strawberry quality. They showed that genotypes affected firmness, anthocyanin content, and antioxidant capacity, while harvest date contributed to the appearance, color, TSS, and TA, and their interactions affected total phenolic content.

The brief report by Hartman et al. (contribution 10) analyzed the effect of water and fertilizer on two Juneberry cultivars and seedling windbreak plantings. Natural conditions (no water or fertilizers) were compared with irrigation during flowering and irrigation plus fertilization during flowering and fruit ripening periods during a three-year experiment. Yield varied significantly among the treatments and years. Fertilization increased yield compared to irrigation. Seedlings were recommended for establishing new plantations, and irrigation and fertilization were suggested to increase the yield.

3. Conclusions

This compilation of articles demonstrates the significant effects of genotype and the environment on fruit production and quality. It emphasizes that more than one growing season is needed to see the beneficial effects of organic matter in improving the yield and fruit quality. It also emphasizes the importance of mineral fertilization, which may not be entirely replaced by organic fertilization without reducing yield. However, the fruit quality may be superior with organic fertilizers compared to that grown with mineral fertilization.

As humans prepare to feed a larger population, extreme climate events that are detrimental to agriculture production systems become more frequent and widespread, and humans start investing in the growing plants in outer space for extraterrestrial trips, we may need to invest in innovative technologies to grow high-quality and nutritious plants in controlled environments. The application of developing techniques and new technologies, such as machine learning, to help us identify defects is also emphasized. Emerging technologies, such as artificial intelligence, big data, the Internet of Things, and cloud computing, will play an important role in our mission to feed human populations, and need to be further investigated [5].

The other arena is via breeding plants that are more resistant to the extreme conditions they may be exposed to. The information in this Special Issue will help us to better understand small fruit production systems. It may contribute to the breeding of cultivars with attractive fruits of higher quality and nutritional value.

Conflicts of Interest: The author declares no conflicts of interest.

List of Contributions

1. Xie, Y.; Black, Z.; Xu, N.; Brecht, J.; Huff, D.; Zhao, X. Influence of Sunn Hemp Biomass Incorporation on Organic Strawberry Production. *Horticulturae* **2023**, *9*, 1247. https://doi.org/10.3390/horticulturae9111247.
2. Patel, H.; Taghavi, T.; Samtani, J. Fruit Quality of Several Strawberry Cultivars during the Harvest Season under High Tunnel and Open Field Environments. *Horticulturae* **2023**, *9*, 1084. https://doi.org/10.3390/horticulturae9101084.
3. Marcellini, M.; Raffaelli, D.; Mazzoni, L.; Pergolotti, V.; Balducci, F.; Armas Diaz, Y.; Mezzetti, B.; Capocasa, F. Effects of Different Irrigation Rates on Remontant Strawberry Cultivars Grown in Soil. *Horticulturae* **2023**, *9*, 1026. https://doi.org/10.3390/horticulturae9091026.
4. Čabilovski, R.; Manojlović, M.; Popović, B.; Radojčin, M.; Magazin, N.; Petković, K.; Kovačević, D.; Lakićević, M. Vermicompost and Vermicompost Leachate Application in Strawberry Production: Impact on Yield and Fruit Quality. *Horticulturae* **2023**, *9*, 337. https://doi.org/10.3390/horticulturae9030337.
5. Mbarushimana, J.; Bosch, D.; Samtani, J. An Economic Comparison of High Tunnel and Open-Field Strawberry Production in Southeastern Virginia. *Horticulturae* **2022**, *8*, 1139. https://doi.org/10.3390/horticulturae8121139.
6. Hosainpour, A.; Kheiralipour, K.; Nadimi, M.; Paliwal, J. Quality Assessment of Dried White Mulberry (*Morus alba* L.) Using Machine Vision. *Horticulturae* **2022**, *8*, 1011. https://doi.org/10.3390/horticulturae8111011.
7. Zydlik, Z.; Zydlik, P.; Kafkas, N.; Yesil, B.; Cieśliński, S. Foliar Application of Some Macronutrients and Micronutrients Improves Yield and Fruit Quality of Highbush Blueberry (*Vaccinium corymbosum* L.). *Horticulturae* **2022**, *8*, 664. https://doi.org/10.3390/horticulturae8070664.
8. Rivero, R.; Remberg, S.; Heide, O.; Sønsteby, A. Effect of Temperature and Photoperiod Pre-conditioning on Flowering and Yield Performance of Three Everbearing Strawberry Cultivars. *Horticulturae* **2022**, *8*, 504. https://doi.org/10.3390/horticulturae8060504.
9. Zhang, Y.; Yang, M.; Hou, G.; Zhang, Y.; Chen, Q.; Lin, Y.; Li, M.; Wang, Y.; He, W.; Wang, X.; Tang, H.; Luo, Y. Effect of Genotype and Harvest Date on Fruit Quality, Bioactive Compounds, and Antioxidant Capacity of Strawberry. *Horticulturae* **2022**, *8*, 348. https://doi.org/10.3390/horticulturae8040348.
10. Hartman, K.; Alahakoon, D.; Fennell, A. Impact of Water and Nutrient Supplementation on Yield of Prairie Plantings of Juneberry Amelanchier alnifolia Nutt., Cultivar and Windbreak Plantings. *Horticulturae* **2023**, *9*, 653. https://doi.org/10.3390/horticulturae9060653.

References

1. Koukounaras, A. Advanced Greenhouse Horticulture: New Technologies and Cultivation Practices. *Horticulturae* **2021**, *7*, 1. [CrossRef]
2. Xu, X. Major challenges facing the commercial horticulture. *Front. Hortic.* **2022**, *1*, 980159. [CrossRef]
3. Lamm, K.; Powell, A.; Lombardini, L. Identifying Critical Issues in the Horticulture Industry: A Delphi Analysis during the COVID-19 Pandemic. *Horticulturae* **2021**, *7*, 416. [CrossRef]

4. Ludwig-Ohm, S.; Hildner, P.; Isaak, M.; Dirksmeyer, W.; Schattenberg, J. The contribution of Horticulture 4.0 innovations to more sustainable horticulture. *Procedia Comput. Sci.* **2023**, *217*, 465–477. [CrossRef]
5. Zhang, M.; Han, Y.; Li, D.; Xu, S.; Huang, Y. Smart horticulture as an emerging interdisciplinary field combining novel solutions: Past development, current challenges, and future perspectives. *Hortic. Plant J.* **2023**. [CrossRef]

 *horticulturae*

Article

Effect of Genotype and Harvest Date on Fruit Quality, Bioactive Compounds, and Antioxidant Capacity of Strawberry

Yunting Zhang [1,2,†], Min Yang [1,†], Guoyan Hou [1], Yong Zhang [1], Qing Chen [1], Yuanxiu Lin [1,2], Mengyao Li [1], Yan Wang [1,2], Wen He [1,2], Xiaorong Wang [1,2], Haoru Tang [1,2,*] and Ya Luo [1,*]

[1] College of Horticulture, Sichuan Agricultural University, Chengdu 611130, China; asyunting@gmail.com (Y.Z.); ym2022215@163.com (M.Y.); 2021205002@stu.sicau.edu.cn (G.H.); zhyong@sicau.edu.cn (Y.Z.); supnovel@gmail.com (Q.C.); linyx@sicau.edu.cn (Y.L.); limy@sicau.edu.cn (M.L.); wangyanwxy@163.com (Y.W.); hewen0724@gmail.com (W.H.); wangxr@sicau.edu.cn (X.W.)

[2] Institute of Pomology and Olericulture, Sichuan Agricultural University, Chengdu 611130, China

[*] Correspondence: htang@sicau.edu.cn (H.T.); luoya945@163.com (Y.L.)

[†] These authors contributed equally to this work.

Abstract: Fruit quality is strongly affected by genotype and harvest date. In this study, parameters regarding fruit quality, bioactive compounds, and antioxidant capacity of different strawberry cultivars at three harvesting dates were quantified to elucidate the influence of genotype and harvest date on strawberry quality. The results showed that harvest date was the major contributor to appearance color, TSS, TA, and TSS/TA ratio of strawberries, while genotype mainly affected firmness, anthocyanin content, and antioxidant capacity. Moreover, the interaction of genotype and harvest date had a primary influence on the content of ascorbic acid. The content of total phenolics and amino acids received the similar influence caused by genotype and harvest date. However, the interaction of genotype and harvest date significantly affected total phenolic content as well. These findings give a better understanding of the influence of the genotype and harvest date on strawberry, which might contribute to breed cultivars with more attractive fruits in terms of quality acceptance and nutritional value.

Keywords: strawberry; genotype; harvest date; fruit quality; amino acid

1. Introduction

Strawberry (*Fragaria* × *ananassa* Duch.) is one of the most popular and commonly consumed small fruits worldwide. According to the Food and Agriculture Organization (FAO) of the United Nations, world production of strawberries exceeds eight million tons (FAO 2018). In addition to its unique flavor, attractive color, and preferred organoleptic properties, strawberry is particularly a rich source of a wide variety of nutritive and non-nutritive bioactive compounds, which exert a synergistic and cumulative effect on human health promotion and on the prevention of various diseases such as cancer, cardiovascular diseases, obesity, diabetes, inflammation, and neurological diseases [1–3]. Hence, the development of strawberry cultivars rich in bioactive compounds and better flavor is the objective of breeding.

Fruit flavor mostly depends on the contents of total soluble solids (TSS), total acids (TA), and their ratio [4]. TSS is a group of substances that can be dissolved in water, containing sugar, acid, vitamins, minerals, and so on, and sugar is a major constituent of total soluble solids [5,6]. It has been recognized that free amino acids also affect fruit flavor, since some of them are aroma precursors and taste key determinants during fruit maturation [7]. Strawberry fruits are good sources of various kinds of amino acids [8]. The ripe strawberry color is mainly triggered by the composition and contents of anthocyanins [9]. The bioactive compounds in strawberries such as ascorbic acid and polyphenol present high antioxidant capacity [10]. It has been documented that genotypic variation mainly

Citation: Zhang, Y.; Yang, M.; Hou, G.; Zhang, Y.; Chen, Q.; Lin, Y.; Li, M.; Wang, Y.; He, W.; Wang, X.; et al. Effect of Genotype and Harvest Date on Fruit Quality, Bioactive Compounds, and Antioxidant Capacity of Strawberry. *Horticulturae* **2022**, *8*, 348. https://doi.org/10.3390/horticulturae8040348

Academic Editor: Toktam Taghavi

Received: 13 February 2022
Accepted: 14 April 2022
Published: 18 April 2022

Publisher's Note: MDPI stays neutral with regard to jurisdictional claims in published maps and institutional affiliations.

determined these attributes, while harvest date was also an important factor influencing the chemical composition of strawberries [11–15]. For instance, Samykanno et al. [12] reported that both genotype and harvest date had a significant effect on pH, titratable acidity (TA), and TSS/TA ratio, but variations of pH and titratable acidity (TA) in strawberry fruits were mainly attributed to genotypic differences, while total soluble solid (TSS) content was largely related to harvest date. Chandler et al. [16] noted that cultivar rankings shifted for TSS, but not for TA between harvest dates. Winardiantika et al. [11] showed that the total anthocyanin and phenolic compounds received a significant synergistic influence from harvest date and strawberry genotype, which may give rise to the variation of antioxidant capacity [13].

Given the effects of genotype and harvest date on strawberry characteristics, the aim of this study was to evaluate fruit quality, bioactive compounds, and antioxidant capacity of three cultivars harvested at three different times, as well as to propose the genotypes and the harvest date to obtain better quality fruits.

2. Materials and Methods

2.1. Experimental site and Plant Materials

Three strawberry cultivars including 'Kaolino', 'Benihoppe', and 'Hongyu' (Figure 1) were grown in a plastic greenhouse under natural conditions in a commercial farm (Dr. Luo Agricultural Development Co., Ltd., Ya'an, China). Strawberry planting was accomplished on 20 August 2018. Fruits were harvested at commercial maturity (three-fourths to full red) on 20 January, 20 February, and 20 March 2019, respectively. These materials were quickly transferred to laboratory. Some samples were used to assess fruit surface color and firmness, and others were frozen in liquid nitrogen and then stored at −80 °C for nutrient composition and antioxidant capacity analysis.

Figure 1. The ripe fruits of three cultivars.

2.2. Fruit Color and Firmness Assessment

The fruit color assessment was performed using a handheld chromameter CR-400 (Konica Minolta, Japan) determining L^*, a^*, b^* color parameters. The L^* scale ranges from 0 (the darkest black) to 100 (the brightest white) to represent lightness. The a^* and b^* scales indicate color composition; a^* is the scale of green to red from negative to positive direction, whereas b^* varies from blue to yellow. The fruit firmness was measured with a fruit firmness meter (FR-5105, Lutron, Japan). Each fruit was measured twice in opposite sides of its equatorial zone [17]. Results were recorded in newtons (N).

2.3. Total Soluble Solid, Titratable Acid, and Ascorbic Acid Content Determination

The total soluble solid (TSS) content of the squeezed strawberry juice was measured using a pocket refractometer (PAL-1, Atago, Japan) and expressed as a percentage (%). The titratable acid (TA) content was determined by titration method with 0.1 mol/L NaOH and expressed as the percentage of citric acid on a fresh weight (FW) basis [18]. The ascorbic acid (AsA) was determined according to the method of Zhang and Kirkham [19] and expressed as mg of AsA per 100 g of FW.

2.4. Total Anthocyanin, Phenolic Content, and Antioxidant Capacity Assay

Total anthocyanins were determined by a pH differential method at 496 nm and 700 nm in buffers at pH 1.0 and 4.5 [20] and expressed as mg of pelargonidin-3-glucoside per 100 g FW. Total phenolic content was estimated according to the Folin–Ciocâlteu procedure as used by Molan et al. [21]. Briefly, the extract was mixed with 2% sodium carbonate solution for 5 min. Then, Folin–Ciocâlteu phenol reagent (50%) was added and allowed to stand for 30 min. All reactions were conducted at room temperature. The absorbance was read at 650 nm with a plate reader (Varioskan LUX, Thermo Fisher Scientific, Waltham, MA, USA). Gallic acid (Sangon, Shanghai, China) was used as a standard, and the results were expressed as mg of gallic acid per g FW.

The 1,1-diphenyl-2-picryl-hydrazyl (DPPH) radical-scavenging capacity and the ferric reducing antioxidant power (FRAP) assays were used to evaluate strawberry antioxidant capacity. DPPH assay was conducted according to the method outlined by Brand-Williams et al. [22] with some modifications. Briefly, 2.8 mL of a 60 μM solution of DPPH in ethanol was mixed with 200 μL of sample solution. The mixture in the test tubes was shaken well and incubated in the dark for 30 min at room temperature. Then, the absorbance was recorded at 517 nm. The scavenging activity was estimated according to the inhibition percentage of DPPH using the following equation: % inhibition of DPPH radical activity = (absorbance $_{Control}$ − absorbance $_{sample}$)/absorbance $_{Control}$ × 100. The FRAP assay was performed according to the method described by Benzie and Strain [23]. Briefly, 20 μL of extract was added to 1.8 mL of the working FRAP reagent consisting of 300 mM acetate buffer (pH 3.6), 10 mM TPTZ, and 20 mM $FeCl_3 \cdot 6H_2O$ in a 10:1:1 ($v/v/v$) ratio. Then, the reaction mixture was warmed at 37 °C for 30 min before use. The absorbance was measured at 593 nm. The results were expressed as mmol $FeSO_4 \cdot 7H_2O$ per 100 g of FW.

2.5. Amino Acid Content Measurement

Amino acids were detected by an amino-acid automatic analyzer (Hitachi L8900, Tokyo, Japan) according to the previous method [24]. Different amino acids were eluted on a sulfonic acid cation exchange column (4.6 mm × 60 mm, 3 μm) using several buffers with different pH and ion concentrations at 0.04 mL/min sequentially. The ninhydrin flow rate was kept at 0.35 mL/min. The reaction and separation columns were set at 135 °C and 57 °C, and channels 1 and 2 were set at 570 nm and 440 nm, respectively. The injection volume was 20 μL. The amino-acid standards were purchased from Sinopharm Chemical Reagent Co., Ltd.

2.6. Statistical Analysis

Data were subjected to analysis of variance using the general linear model procedure to determine the main effects and interactions, and means were compared using Duncan's multiple range test at a significance level of 0.05. Moreover, principal component analysis (PCA) was analyzed using R package 4.1.1; FactoMineR was used to compute PCA, and factoextra was used to produce ggplot2-based visualization of PCA results.

3. Results

3.1. Source of Variation (F Value) in Quality Traits of Strawberry

To analyze the variation source of quality traits of strawberry across harvest dates, statistical F-tests were conducted. As shown in Table 1, both genotype and harvest date could significantly cause the variation in quality attributes except FRAP, while the interaction of genotype and harvest date significantly gave rise to the variation of multiple quality parameters, except antioxidant activities and TAA level. It also appears that the variation in L^*, a^*, TSS, TA, and TSS/TA ratio was mainly attributed to harvest date, while the variation in F, b^*, ANT, and DPPH was primarily triggered by genotype, and only the variation in AsA was mainly due to genotype and harvest date interaction. In addition, the effect of genotype, harvest date, and their interaction on TP was similar, whereas genotype and harvest date had a similar effect on TAA.

Table 1. Source of variation (F-value) in quality traits of strawberry.

Quality Trait	F-Value (Significance)		
	Harvest Date	**Cultivar**	**Harvest Date × Cultivar**
F	21.183 **	40.416 **	17.339 **
L^*	156.776 **	125.703 **	4.576 **
b^*	16.537 **	89.085 **	12.414 **
a^*	80.503 **	32.401 **	19.431 **
TSS	110.55 **	20.098 **	5.062 **
TA	692.841 **	183.217 **	170.539 **
TSS/TA	692.841 **	183.217 **	170.539 **
AsA	39.143 **	44.581 **	141.297 **
TP	6.942 **	5.719 *	4.457 *
ANT	19.979 **	274.147 **	12.206 **
DPPH	6.635 **	11.879 **	1.212
FRAP	2.634	2.947	0.396
TAA	12.073 **	12.328 **	0.502

*, ** F-values are significant at the 95% and 99% confidence interval, respectively. F, firmness; TSS, total soluble solids; TA, titratable acidity; AsA, ascorbic acid; TP, total phenolics; ANT, anthocyanin; DPPH, 1,1-diphenyl-2-picryl-hydrazyl; FRAP, ferric reducing antioxidant power; TAA, total amino acids.

3.2. Fruit Firmness and Surface Color

Fruit firmness and color are important factors that can affect consumers' quality perception of strawberry fruit. Firmness in cultivars 'Benihoppe' and 'Hongyu' had no significant differences and showed a similar trend at different harvest dates. Cultivar 'Kaolino' showed lower firmness than the other two analyzed cultivars, especially in March. The value of L^* decreased in three cultivars at the later harvesting date. 'Benihoppe' had a significantly lower L^* value at each harvest date, compared with 'Kaolino' and 'Hongyu'. Value a^* was generally higher in February strawberries. Furthermore, cultivar 'Kaolino' showed a relatively higher b^* value than other two cultivars over the harvest date (Figure 2).

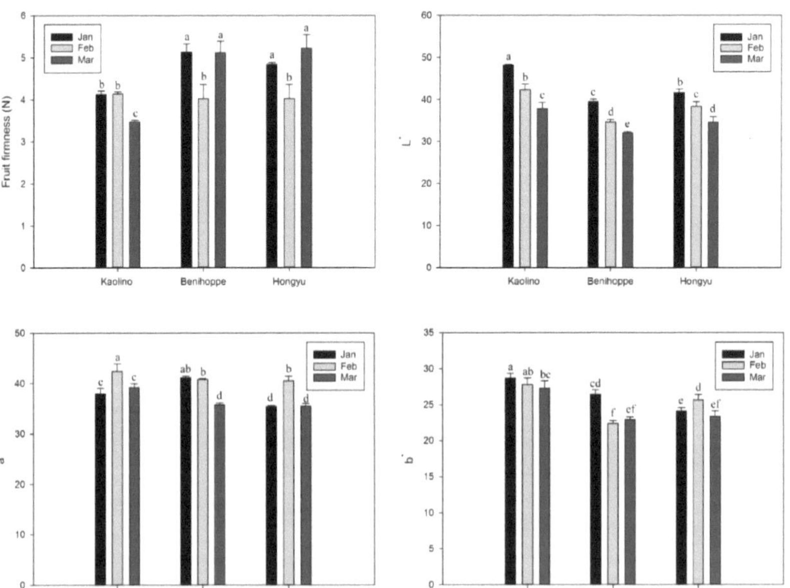

Figure 2. Firmness and appearance color in three cultivars of strawberry fruits at three harvest dates. Values represent the mean ± standard error. Different lowercase letters indicate statistically significant differences at $p \leq 0.05$ as determined by Duncan's test.

3.3. TSS, TA, TSS/TA Ratio, AsA, Total Phenolics, and Anthocyanins

All genotypes had a higher TSS content in January and February than in March, especially for Benihoppe. There was a great seasonal variation for TA content, and the highest TA content was obtained in January in three cultivars. The TSS/TA ratio in 'Kaolino' or 'Hongyu' was highest in February and showed no significant difference between January and March. In 'Benihoppe', the TSS/TA ratio remarkably increased at the later harvesting date. The highest and lowest contents of AsA were obtained in January and February, respectively for Benihoppe. AsA content in 'Hongyu' noticeably increased at the later harvesting date, while AsA content in 'Kaolino' showed no significant difference in January and March. Clearly, harvest date did not affect the accumulation of total phenolics and anthocyanins in 'Benihoppe'. The total phenolic content in 'Benihoppe' from harvest date to harvest date was generally lower than the other two cultivars. 'Kaolino' had lower anthocyanin content than 'Benihoppe' and 'Hongyu' for the three harvest dates, while 'Benihoppe' accumulated more anthocyanins than 'Hongyu' except for March (Figure 3).

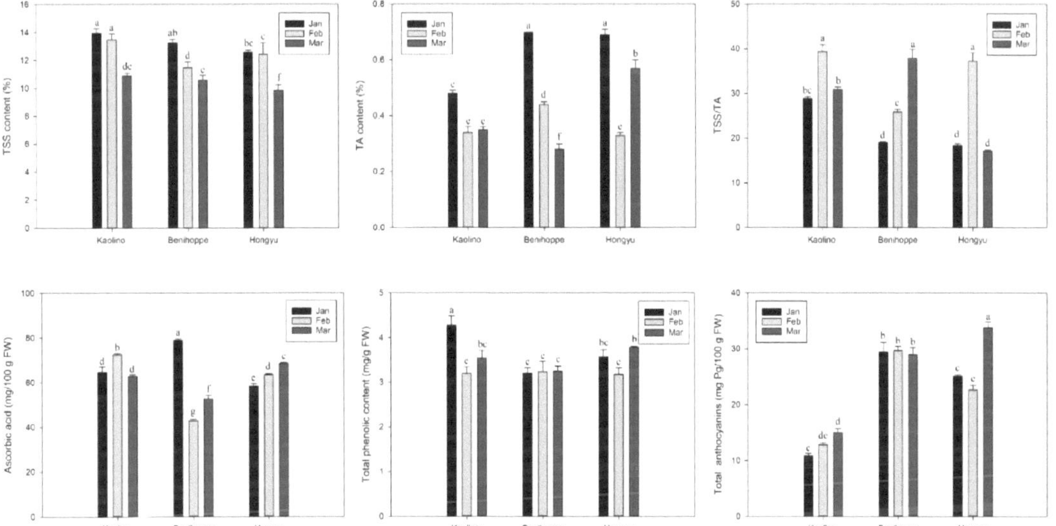

Figure 3. Total soluble solids (TSS), titratable acidity (TA), TSS/TA ratio, ascorbic acid (AsA), total phenolics (TP), and anthocyanins (ANT) in three cultivars of strawberry fruits at three harvest dates. Values represent the mean ± standard error. Different lowercase letters indicate statistically significant differences at $p \leq 0.05$ as determined by Duncan's test.

3.4. Antioxidant Capacity

The antioxidant capacity was investigated using both FRAP and DPPH assays. The lowest antioxidant capacity was detected in February, but showed almost no significant difference from the other two harvesting dates in the same cultivar. The three cultivars were not significantly different at the three harvesting dates at the FRAP level. Interestingly, the antioxidant capacity of 'Kaolino' was lower than the other two cultivars at all harvesting dates except February 'Benihoppe' at the DPPH level (Figure 4).

3.5. Amino Acid

The amino acids of the three cultivars at three different harvesting dates are shown in Tables 2 and 3. A total of 15 amino acids could be detected in this study. Of the eight essential amino acids, strawberries had five (Lys, Phe, Thr, Leu, and Val). Asp and Glu were the main amino acids found in the three cultivars. Their content declined at the later harvesting date, but was hardly affected by harvest date within certain cultivars. In March,

'Kaolino' had the highest levels of Thr and Ser, compared with 'Benihoppe' and 'Hongyu'. The content of Gly in 'Kaolino' at all three harvesting stages was higher than the other two cultivars except for 'Benihoppe' in March. There was no significant difference for Ala among all genotypes at the same harvest date. Cys was almost undetectable in all samples. Harvest date and genotype had almost no impact on Ser, Val, and Phe. Leu, Arg, and Pro showed the highest levels in January for each cultivar, with almost no significant difference between February and March among the three cultivars. The change in harvest date did not obviously influence the Tyr accumulation in a specific cultivar. Lys content in January was highest in 'Benihoppe' and 'Hongyu', but the opposite result in 'Kaolino'. The pattern of His accumulation was almost consistent in all cultivars across all harvesting dates. Obviously, the total amino acids (TAA) in 'Kaolino' were highest on all three harvest dates, while the TAA in 'Benihoppe' was lowest. Additionally, the TAA in 'Benihoppe' and 'Hongyu' gradually decreased at the later harvesting date.

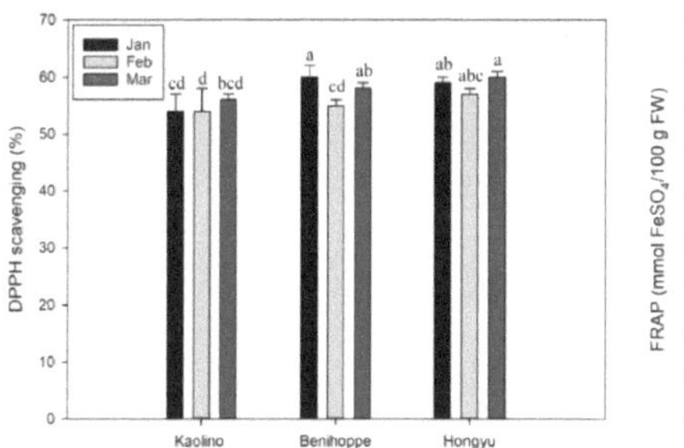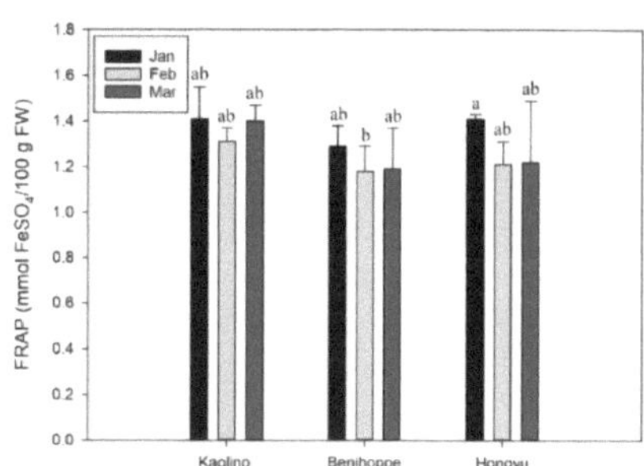

Figure 4. Antioxidant capacity in three cultivars of strawberry fruits at three harvest dates measured by DPPH and FRAP. Values represent the mean ± standard error. Different lowercase letters indicate statistically significant differences at $p \leq 0.05$ as determined by Duncan's test.

Table 2. Amino acid content in three cultivars of strawberry fruits at three harvest dates.

Harvest Date	Cultivar	Asp	Thr	Ser	Glu	Ala	Gly	Cys	Val
January	Kaolino	1.06 ± 0.09 a	0.15 ± 0.00 b	0.18 ± 0.02 b	1.48 ± 0.21 a	0.18 ± 0.01 a	0.20 ± 0.02 abcd	0.01 ± 0.00 a	0.11 ± 0.01 b
	Benihoppe	0.91 ± 0.04 ab	0.11 ± 0.01 c	0.12 ± 0.01 b	1.19 ± 0.09 ab	0.11 ± 0.00 cd	0.20 ± 0.00 abc	0.00 ± 0.00 c	0.16 ± 0.01 a
	Hongyu	1.05 ± 0.10 a	0.11 ± 0.00 c	0.17 ± 0.02 b	1.37 ± 0.09 ab	0.11 ± 0.00 bc	0.21 ± 0.00 a	0.00 ± 0.00 c	0.15 ± 0.02 a
February	Kaolino	1.00 ± 0.02 ab	0.10 ± 0.00 c	0.12 ± 0.01 b	1.33 ± 0.05 ab	0.13 ± 0.01 b	0.18 ± 0.00 cd	0.00 ± 0.00 c	0.11 ± 0.01 b
	Benihoppe	0.86 ± 0.06 b	0.15 ± 0.01 b	0.13 ± 0.01 b	1.06 ± 0.10 b	0.07 ± 0.01 e	0.18 ± 0.01 d	0.01 ± 0.00 ab	0.10 ± 0.01 b
	Hongyu	0.95 ± 0.03 ab	0.10 ± 0.00 c	0.13 ± 0.01 b	1.14 ± 0.08 b	0.10 ± 0.00 cd	0.19 ± 0.00 abcd	0.00 ± 0.00 c	0.13 ± 0.00 ab
March	Kaolino	0.85 ± 0.05 b	0.30 ± 0.01 a	0.26 ± 0.06 a	1.18 ± 0.02 ab	0.17 ± 0.01 a	0.20 ± 0.00 ab	0.00 ± 0.00 bc	0.12 ± 0.01 ab
	Benihoppe	0.84 ± 0.03 b	0.11 ± 0.01 c	0.14 ± 0.02 b	0.74 ± 0.07 c	0.19 ± 0.00 a	0.18 ± 0.00 bcd	0.01 ± 0.00 a	0.11 ± 0.00 b
	Hongyu	0.88 ± 0.04 ab	0.12 ± 0.00 c	0.12 ± 0.00 b	1.05 ± 0.07 b	0.09 ± 0.00 de	0.20 ± 0.01 ab	0.00 ± 0.00 c	0.13 ± 0.01 ab

Values represent the mean ± standard error. Different lowercase letters in the same column indicate statistically significant differences at $p \leq 0.05$ as determined by Duncan's test.

Table 3. Amino acid content in three cultivars of strawberry fruits at three harvest dates.

Harvest Date	Cultivar	Leu	Tyr	Phe	Lys	His	Arg	pro	Total AA
January	Kaolino	0.11 ± 0.00 b	0.07 ± 0.00 a	0.18 ± 0.01 a	0.11 ± 0.00 e	0.07 ± 0.00 a	0.12 ± 0.00 a	0.11 ± 0.00 a	4.13 ± 0.35 a
	Benihoppe	0.16 ± 0.01 a	0.06 ± 0.01 ab	0.17 ± 0.02 a	0.13 ± 0.01 bc	0.06 ± 0.00 a	0.11 ± 0.01 abc	0.08 ± 0.00 bc	3.57 ± 0.18 bcd
	Hongyu	0.11 ± 0.01 b	0.05 ± 0.00 bc	0.15 ± 0.01 ab	0.15 ± 0.00 a	0.07 ± 0.00 a	0.12 ± 0.00 ab	0.10 ± 0.01 a	3.92 ± 0.15 ab
February	Kaolino	0.08 ± 0.00 cd	0.06 ± 0.00 ab	0.17 ± 0.02 a	0.12 ± 0.00 cd	0.05 ± 0.00 b	0.08 ± 0.00 de	0.08 ± 0.00 bc	3.59 ± 0.07 bc
	Benihoppe	0.06 ± 0.00 d	0.04 ± 0.00 c	0.17 ± 0.01 a	0.11 ± 0.00 e	0.05 ± 0.00 b	0.06 ± 0.00 e	0.00 ± 0.00 d	3.06 ± 0.13 de
	Hongyu	0.08 ± 0.00 cd	0.04 ± 0.00 c	0.12 ± 0.00 b	0.07 ± 0.00 f	0.05 ± 0.00 b	0.09 ± 0.01 cd	0.07 ± 0.00 c	3.26 ± 0.11 cde
March	Kaolino	0.09 ± 0.00 bc	0.07 ± 0.01 a	0.18 ± 0.01 a	0.14 ± 0.00 b	0.06 ± 0.00 a	0.10 ± 0.01 bcd	0.08 ± 0.00 b	3.81 ± 0.04 ab
	Benihoppe	0.07 ± 0.01 cd	0.07 ± 0.00 a	0.18 ± 0.00 a	0.11 ± 0.00 de	0.03 ± 0.00 c	0.09 ± 0.01 cd	0.07 ± 0.00 c	2.94 ± 0.13 e
	Hongyu	0.08 ± 0.00 cd	0.05 ± 0.00 bc	0.15 ± 0.00 ab	0.12 ± 0.00 de	0.06 ± 0.00 a	0.08 ± 0.00 de	0.07 ± 0.00 c	3.21 ± 0.08 cde

Values represent the mean ± standard error. Different lowercase letters in the same column indicate statistically significant differences at $p \leq 0.05$ as determined by Duncan's test.

Principal component analysis (PCA) was used for the investigation of necessary components to explain the greater part of variance with a minimum loss of information and to find the relationship between objects. The results explained that the first principal component (PC1) and PC2 accounted for 42.1% and 21.7% of the total variation, respectively. As shown in Figure 5, most of amino acids were grouped on the right side of PCA plot. By overlapping the positions of amino acids and analyzed samples, it was evident that 'Kaolino' and strawberries produced in January showed higher performance in terms of amino acids, indicating that the content of amino acids was co-regulated by genotype and harvest date.

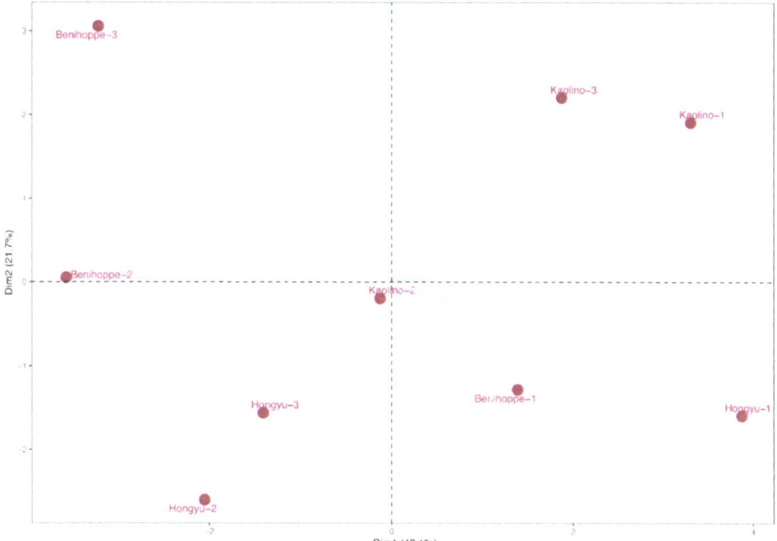

Figure 5. Principal component analysis (PCA) showing interrelation of amino acids with the combination of cultivar and harvest date.

4. Discussion

The quality of strawberry is a result of the complex balance among appearance, aroma, texture, and sweetness, which affects consumers' liking and willingness. In recent years, the high content of health-promoting compounds such as anthocyanin, ascorbic acid, and polyphenol in strawberry has attracted more attention [25–27]. Several reports available in the literature indicate that the key factors affecting fruit quality attributes and bioactive compounds include genotype and harvest dates [28–30].

It is well known that different strawberry varieties display a wide range of fruit firmness [17,31,32]. In the present study, 'Benihoppe' and 'Hongyu' fruits were clearly firmer than 'Kaolino', although the firmness was affected by harvest date. The analysis of variance showed that the variation in firmness could be mainly attributed to genotypic differences since relatively large differences among the firmness of the three cultivars were observed for three harvest dates. Hence, it is a potential strategy to conduct strawberry classification with respect to firmness. Šamec et al. [13] reported that appearance color was more under the influence of genotype, whereas sampling dates did not affect the color of the cultivars. In contrast with that study, our results showed that both harvest date and genotype had a significant effect on color (L^* and a^* value), with harvest date playing a greater role. The value of L^* decreased in 'Kaolino', 'Benihoppe', and 'Hongyu' at the later harvesting date, while the value of a^* was generally higher in February strawberries.

In addition to firmness and appearance color, parameters such as TA, TSS, and TSS/TA have a critical role in determining strawberry fruit quality. In our study, TA, TSS and

TSS/TA ratio depended on harvest date and genotype, consistent with the observations of other authors [12,33]. Moreover, their variance caused by harvest date appeared to be larger. However, some other reports showed that the variance of TA was mainly genetically determined, while the variance of TSS may have been predominantly due to the harvest date [12,16,34,35]. 'Kaolino' and 'Hongyu' were interesting for their higher TSS/TA ratio in February, while 'Benihoppe' had a higher TSS/TA ratio in March. For an acceptable strawberry flavor, a minimum TSS of 7% and a maximum TA of 0.8% are recommended [36]. Hence, continuously monitoring TSS, TA, and TSS/TA ratio across production season can individuate the optimum strawberry harvesting time for better taste quality to fulfill consumers' expectations.

AsA, TP, and ANT in fruit play important roles in scavenging reactive oxygen species. It has been documented that they are controlled by genotype and external conditions [37–39]. Our results showed that cultivar and harvest date noticeably influenced their accumulation, and significant interactions between genotype and harvest date was also observed in these traits, in accordance with a previous study in strawberry [40]. Anthocyanin was more affected by genotype, as reported elsewhere [39,41], while AsA was more affected by the interaction of genotype and harvest date. In addition, FRAP data showed that the antioxidant capacity was not influenced by cultivars and harvest date, in contrast to the result of DPPH, which was significantly regulated by the two factors. Clearly, the antioxidant capacity of 'Kaolino' was lower than that of the other two cultivars across all harvesting dates except 'Benihoppe' in February at the DPPH level. The different results of antioxidant capacity given by the FRAP and DPPH method may be due to their different reaction principles.

Several studies have shown that amino acids influence plant resistance and fruit quality [42–44]. For instance, proline is involved in response to various environmental signals related to abiotic or biotic stress [45–47]. L-Glutamate is responsible for 'umami' or delicious taste; it was reported that glutamate is the principal free amino acid of ripe fruits of cultivated varieties in tomato [42]. L-Glycine, L-alanine, serine, and proline provide sweetness [7]. Hence, an investigation of the factors influencing amino-acid accumulation can contribute to strawberry breeding. It was shown that Asp and Glu were the main amino acids found in 'Kaolino', 'Benihoppe', and 'Hongyu', while different results were obtained in other strawberry cultivars [8,45]. The contents of Asp and Glu decreased at the later harvesting date, but were hardly affected by harvest date within a certain cultivar. However, the total amino acid (TAA) displayed a similar effect of genotype and harvest date. 'Kaolino' had the highest TAA across the three harvest dates, while 'Benihoppe' had the lowest TAA content. The TAA content gradually declined in 'Benihoppe' and 'Hongyu' at the later harvesting date. PCA results confirmed that the discrepancy in amino acids of strawberry was related to cultivar differences and seasonal collection.

5. Conclusions

The fruit quality, bioactive compounds, and antioxidant capacity of strawberries were jointly regulated by genotype and harvest date. In particular, the harvest date was closely influenced by environmental factors such as light, temperature, and water. Consequently, growers may make an effort to manage environmental factors during fruit production to achieve stability of the above properties after selecting a suitable cultivar.

Author Contributions: Conceptualization, H.T. and Y.L. (Ya Luo); methodology, Y.Z. (Yunting Zhang) and M.Y.; validation, M.Y. and G.H.; visualization, Y.Z. (Yong Zhang) and Q.C.; software, Y.L. (Yuanxiu Lin) and M.L.; formal analysis, Y.W. and W.H.; writing—original draft preparation, Y.Z. (Yunting Zhang); writing—review and editing, Y.L. (Ya Luo); supervision, X.W., H.T. and Y.L. (Ya Luo). All authors have read and agreed to the published version of the manuscript.

Funding: This study was supported by the National Natural Science Foundation of China (3180817), the State Education Ministry, Key projects of Sichuan Provincial Education Department (172A0319) and Key projects of Sichuan Provincial Science and Technology Department (2018NZ0126), and the Service Station Projects of New Rural Development Research Institute, Sichuan Agricultural University (2018, 2020).

Institutional Review Board Statement: Not applicable.

Informed Consent Statement: Not applicable.

Data Availability Statement: Data are contained within this article.

Acknowledgments: The authors thank Yuntian Ye for the technical assistance with principal component analysis.

Conflicts of Interest: The authors declare no conflict of interest.

References

1. Afrin, S.; Gasparrini, M.; Forbes-Hernandez, T.Y.; Reboredo-Rodriguez, P.; Mezzetti, B.; Varela-López, A.; Giampieri, F.; Battino, M. Promising health benefits of the strawberry: A focus on clinical studies. *J. Agric. Food Chem.* **2016**, *64*, 4435–4449. [CrossRef] [PubMed]
2. Giampieri, F.; Forbes-Hernandez, T.Y.; Gasparrini, M.; Alvarez-Suarez, J.M.; Afrin, S.; Bompadre, S.; Quiles, J.L.; Mezzetti, B.; Battino, M. Strawberry as a health promoter: An evidence based review. *Food Funct.* **2015**, *6*, 1386–1398. [CrossRef] [PubMed]
3. Battino, M.; Giampieri, F.; Cianciosi, D.; Ansary, J.; Chen, X.; Zhang, D.; Gil, E.; Forbes-Hernández, T. The roles of strawberry and honey phytochemicals on human health: A possible clue on the molecular mechanisms involved in the prevention of oxidative stress and inflammation. *Phytomedicine* **2021**, *86*, 153170. [CrossRef]
4. Krüger, E.; Josuttis, M.; Nestby, R.; Toldam-Andersen, T.; Carlen, C.; Mezzetti, B. Influence of growing conditions at different latitudes of europe on strawberry growth performance, yield and quality. *J. Berry Res.* **2012**, *2*, 143–157. [CrossRef]
5. Huang, X.; Wang, H.; Luo, W.; Xue, S.; Hayat, F.; Gao, Z. Prediction of loquat soluble solids and titratable acid content using fruit mineral elements by artificial neural network and multiple linear regression. *Sci. Hortic.* **2021**, *278*, 109873. [CrossRef]
6. Saridas, M.A.; Simsek, O.; Donmez, D.; Kacar, Y.A.; Kargi, S.P. Genetic diversity and fruit characteristics of new superior hybrid strawberry (*Fragaria× ananassa* Duchesne ex Rozier) genotypes. *Genet. Resour. Crop. Evol.* **2021**, *68*, 741–758. [CrossRef]
7. Ito, H.; Ueno, H.; Kikuzaki, H. Free amino acid compositions for fruits. *J. Nutr. Diet. Pract.* **2017**, *1*, 001–005.
8. Zhang, H.; Wang, Z.-Y.; Yang, X.; Zhao, H.-T.; Zhang, Y.-C.; Dong, A.-J.; Jing, J.; Wang, J. Determination of free amino acids and 18 elements in freeze-dried strawberry and blueberry fruit using an amino acid analyzer and ICP-MS with micro-wave digestion. *Food Chem.* **2014**, *147*, 189–194. [CrossRef]
9. Dzhanfezova, T.; Barba-Espín, G.; Müller, R.; Joernsgaard, B.; Hegelund, J.N.; Madsen, B.; Larsen, D.H.; Vega, M.M.; Toldam-Andersen, T.B. Anthocyanin profile, antioxidant activity and total phenolic content of a strawberry (*Fragaria× ananassa* Duch) genetic resource collection. *Food Biosci.* **2020**, *36*, 100620. [CrossRef]
10. Villamil-Galindo, E.; Van de Velde, F.; Piagentini, A.M. Extracts from strawberry by-products rich in phenolic compounds reduce the activity of apple polyphenol oxidase. *LWT-Food Sci. Technol.* **2020**, *133*, 110097. [CrossRef]
11. Winardiantika, V.; Lee, Y.H.; Park, N.I.; Yeoung, Y.-R. Effects of cultivar and harvest time on the contents of antioxidant phytochemicals in strawberry fruits. *Hortic. Environ. Biotechnol.* **2015**, *56*, 732–739. [CrossRef]
12. Samykanno, K.; Pang, E.; Marriott, P.J. Genotypic and environmental effects on flavor attributes of 'Albion' and 'Juliette' strawberry fruits. *Sci. Hortic.* **2013**, *164*, 633–642. [CrossRef]
13. Šamec, D.; Maretić, M.; Lugarić, I.; Mešić, A.; Salopek-Sondi, B.; Duralija, B. Assessment of the differences in the physical, chemical and phytochemical properties of four strawberry cultivars using principal component analysis. *Food Chem.* **2016**, *194*, 828–834. [CrossRef] [PubMed]
14. Kim, S.K.; Kim, D.S.; Kim, D.Y.; Chun, C. Variation of bioactive compounds content of 14 oriental strawberry cultivars. *Food Chem.* **2015**, *184*, 196–202. [CrossRef]
15. Pelayo-Zaldívar, C.; Ebeler, S.E.; Kader, A.A. Cultivar and harvest date effects on flavor and other quality attributes of california strawberries. *J. Food Qual.* **2005**, *28*, 78–97. [CrossRef]
16. Chandler, C.K.; Herrington, M.; Slade, A. Effect of harvest date on soluble solids and titratable acidity in fruit of strawberry grown in a winter, annual hill production system. *Acta Hortic.* **2003**, *626*, 345–346. [CrossRef]
17. Dotto, M.C.; Martínez, G.A.; Civello, P.M. Expression of expansin genes in strawberry varieties with contrasting fruit firmness. *Plant Physiol. Biochem.* **2006**, *44*, 301–307. [CrossRef]
18. Liu, C.; Zheng, H.; Sheng, K.; Liu, W.; Zheng, L. Effects of melatonin treatment on the postharvest quality of strawberry fruit. *Postharvest Biol. Technol.* **2018**, *139*, 47–55. [CrossRef]
19. Zhang, J.; Kirkham, M. Antioxidant responses to drought in sunflower and sorghum seedlings. *New Phytol.* **1996**, *132*, 361–373. [CrossRef]
20. Zhang, Y.; Hu, W.; Peng, X.; Sun, B.; Wang, X.; Tang, H. Characterization of anthocyanin and proanthocyanidin biosynthesis in two strawberry genotypes during fruit development in response to different light qualities. *J. Photochem. Photobiol. B Biol.* **2018**, *186*, 225–231. [CrossRef]
21. Molan, A.L.; Flanagan, J.; Wei, W.; Moughan, P. Selenium-containing green tea has higher antioxidant and prebiotic activities than regular green tea. *Food Chem.* **2009**, *114*, 829–835. [CrossRef]
22. Brand-Williams, W.; Cuvelier, M.-E.; Berset, C. Use of a free radical method to evaluate antioxidant activity. *LWT-Food Sci. Technol.* **1995**, *28*, 25–30. [CrossRef]

23. Benzie, I.F.; Strain, J.J. The ferric reducing ability of plasma (FRAP) as a measure of "antioxidant power": The FRAP assay. *Anal. Biochem.* **1996**, *239*, 70–76. [CrossRef] [PubMed]

24. Li, M.; Dong, H.; Wu, D.; Chen, H.; Qin, W.; Liu, W.; Yang, W.; Zhang, Q. Nutritional evaluation of whole soybean curd made from different soybean materials based on amino acid profiles. *Food Qual. Saf.* **2020**, *4*, 41–50. [CrossRef]

25. Padula, M.C.; Lepore, L.; Milella, L.; Ovesna, J.; Malafronte, N.; Martelli, G.; de Tommasi, N. Cultivar based selection and genetic analysis of strawberry fruits with high levels of health promoting compounds. *Food Chem.* **2013**, *140*, 639–646. [CrossRef]

26. Szajdek, A.; Borowska, E. Bioactive compounds and health-promoting properties of berry fruits: A review. *Plant Foods Hum. Nutr.* **2008**, *63*, 147–156. [CrossRef]

27. Fecka, I.; Nowicka, A.; Kucharska, A.Z.; Sokół-Łętowska, A. The effect of strawberry ripeness on the content of polyphenols, cinnamates, L-ascorbic and carboxylic acids. *J. Food Compos. Anal.* **2021**, *95*, 103669. [CrossRef]

28. Jouquand, C.; Chandler, C.; Plotto, A.; Goodner, K. A sensory and chemical analysis of fresh strawberries over harvest dates and seasons reveals factors that affect eating quality. *J. Am. Soc. Hortic. Sci.* **2008**, *133*, 859–867. [CrossRef]

29. Hadjipieri, M.; Christofi, M.; Goulas, V.; Manganaris, G.A. The impact of genotype and harvesting day on qualitative attributes, postharvest performance and bioactive content of loquat fruit. *Sci. Hortic.* **2020**, *263*, 108891. [CrossRef]

30. Redpath, L.E.; Gumpertz, M.; Ballington, J.R.; Bassil, N.; Ashrafi, H. Genotype, environment, year, and harvest effects on fruit quality traits of five blueberry (*Vaccinium corymbosum* L.) cultivars. *Agronomy* **2021**, *11*, 1788. [CrossRef]

31. Hirsch, M.; Langer, S.E.; Marina, M.; Rosli, H.G.; Civello, P.M.; Martínez, G.A.; Villarreal, N.M. Expression profiling of endo-xylanases during ripening of strawberry cultivars with contrasting softening rates. Influence of postharvest and hormonal treatments. *J. Sci. Food Agric.* **2021**, *101*, 3676–3684. [CrossRef] [PubMed]

32. Villarreal, N.M.; Rosli, H.G.; Martínez, G.A.; Civello, P.M. Polygalacturonase activity and expression of related genes during ripening of strawberry cultivars with contrasting fruit firmness. *Postharvest Biol. Technol.* **2008**, *47*, 141–150. [CrossRef]

33. Neocleous, D. Effects of cultivars and coco-substrates on soilless strawberry production in cyprus. *J. Berry Res.* **2012**, *2*, 207–213. [CrossRef]

34. Shaw, D. Genotypic variation and genotypic correlations for sugars and organic acids of strawberries. *J. Am. Soc. Hortic. Sci.* **1988**, *113*, 770–774.

35. Shaw, D.V. Response to selection and associated changes in genetic variance for soluble solids and titratable acids contents in strawberries. *J. Am. Soc. Hortic. Sci.* **1990**, *115*, 839–843. [CrossRef]

36. Agulheiro-Santos, A.; Ricardo-Rodrigues, S.; Laranjo, M.; Melgão, C.; Velázquez, R. Non-destructive prediction of total soluble solids in strawberry using near infrared spectroscopy. *J. Sci. Food Agric.* **2022**. [CrossRef]

37. Josuttis, M.; Carlen, C.; Crespo, P.; Nestby, R.; Toldam-Andersen, T.; Dietrich, H.; Krüger, E. A comparison of bioactive compounds of strawberry fruit from europe affected by genotype and latitude. *J. Berry Res.* **2012**, *2*, 73–95. [CrossRef]

38. Viljevac, V.M.; Krunoslav, D.; Ines, M.; Vesna, T.; Dominik, V.; Zvonimir, Z.; Boris, P.; Zorica, J. Season, location and cultivar influence on bioactive compounds of sour cherry fruits. *Plant Soil Environ.* **2017**, *63*, 389–395. [CrossRef]

39. Cocco, C.; Magnani, S.; Maltoni, M.L.; Quacquarelli, I.; Cacchi, M.; Antunes, L.E.C.; D'Antuono, L.F.; Faedi, W.; Baruzzi, G. Effects of site and genotype on strawberry fruits quality traits and bioactive compounds. *J. Berry Res.* **2015**, *5*, 145–155. [CrossRef]

40. Pradas, I.; Medina, J.J.; Ortiz, V.; Moreno-Rojas, J.M. 'Fuentepina'and 'amiga', two new strawberry cultivars: Evaluation of genotype, ripening and seasonal effects on quality characteristics and health-promoting compounds. *J. Berry Res.* **2015**, *5*, 157–171. [CrossRef]

41. Anttonen, M.J.; Hoppula, K.I.; Nestby, R.; Verheul, M.J.; Karjalainen, R.O. Influence of fertilization, mulch color, early forcing, fruit order, planting date, shading, growing environment, and genotype on the contents of selected phenolics in strawberry (*Fragaria* × *ananassa* Duch.) fruits. *J. Agric. Food Chem.* **2006**, *54*, 2614–2620. [CrossRef] [PubMed]

42. Sorrequieta, A.; Ferraro, G.; Boggio, S.B.; Valle, E.M. Free amino acid production during tomato fruit ripening: A focus on l-glutamate. *Amino Acids* **2010**, *38*, 1523–1532. [CrossRef] [PubMed]

43. Sung, J.; Suh, J.H.; Chambers, A.H.; Crane, J.; Wang, Y. Relationship between sensory attributes and chemical composition of different mango cultivars. *J. Agric. Food Chem.* **2019**, *67*, 5177–5188. [CrossRef] [PubMed]

44. Heinemann, B.; Hildebrandt, T.M. The role of amino acid metabolism in signaling and metabolic adaptation to stress-induced energy deficiency in plants. *J. Exp. Bot.* **2021**, *72*, 4634–4645. [CrossRef]

45. Keutgen, A.J.; Pawelzik, E. Contribution of amino acids to strawberry fruit quality and their relevance as stress indicators under nacl salinity. *Food Chem.* **2008**, *111*, 642–647. [CrossRef]

46. Qamar, A.; Mysore, K.; Senthil-Kumar, M. Role of proline and pyrroline-5-carboxylate metabolism in plant defense against invading pathogens. *Front. Plant Sci.* **2015**, *6*, 503. [CrossRef]

47. Batista-Silva, W.; Heinemann, B.; Rugen, N.; Nunes-Nesi, A.; Araújo, W.L.; Braun, H.P.; Hildebrandt, T.M. The role of amino acid metabolism during abiotic stress release. *Plant Cell Environ.* **2019**, *42*, 1630–1644. [CrossRef]

Article

Effect of Temperature and Photoperiod Preconditioning on Flowering and Yield Performance of Three Everbearing Strawberry Cultivars

Rodmar Rivero [1,*], Siv Fagertun Remberg [1], Ola M. Heide [2] and Anita Sønsteby [3]

[1] Faculty of Biosciences, Norwegian University of Life Sciences, NO-1432 Ås, Norway; siv.remberg@nmbu.no
[2] Faculty of Environmental Sciences and Natural Resource Management, Norwegian University of Life Sciences, NO-1432 Ås, Norway; ola.heide@nmbu.no
[3] NIBIO—Norwegian Institute of Bioeconomy Research, NO-1431 Ås, Norway; anita.sonsteby@nibio.no
* Correspondence: rodmar.rivero.casique@nmbu.no; Tel.: +47-90408434

Abstract: Environmental control of flowering in everbearing strawberry is well known, while the optimal commercial raising conditions for high and continuous yield remains unsettled. We exposed freshly rooted plants of cultivars Altess, Favori and Murano to 9 °C, 15 °C, 21 °C and 27 °C, respectively, at two photoperiods for 4 weeks, and assessed flowering and yield performance. Long days at 15–21 °C enhanced flowering, while short days (SD), particularly at 27 °C, decreased flowering. Runner formation was enhanced by SD, being inversely related to flowering. Yields the next season were highest in plants exposed to 15–21 °C, whereas the seasonal harvest distribution varied. In concurrence with earlier reports, the size of the first fruit flush determined the magnitude of the second flush and the length of the off period when little fruit was produced. The large first fruiting flushes of plants exposed to 21 and 27 °C gave particularly long off periods and small second flushes. Moderate first flushes of plants from intermediate temperatures also resulted in a more evenly distributed harvest and the largest yields. Developing flowers and fruits are strong sinks for photosynthates; therefore, the size of the first fruit flush must be compromised to optimize fruit yield and seasonal crop distribution.

Keywords: everbearing; *Fragaria* × *ananassa*; photoperiod; preconditioning; temperature; yield

Citation: Rivero, R.; Remberg, S.F.; Heide, O.M.; Sønsteby, A. Effect of Temperature and Photoperiod Preconditioning on Flowering and Yield Performance of Three Everbearing Strawberry Cultivars. *Horticulturae* **2022**, *8*, 504. https://doi.org/10.3390/horticulturae8060504

Academic Editor: Toktam Taghavi

Received: 6 May 2022
Accepted: 2 June 2022
Published: 6 June 2022

Publisher's Note: MDPI stays neutral with regard to jurisdictional claims in published maps and institutional affiliations.

1. Introduction

In contrast to the traditional seasonal flowering (SF) strawberry cultivars which are quantitative short day (SD) plants at intermediate temperatures (18–21 °C) and unable to flower at high temperature (27 °C), the everbearing (EB) cultivars are quantitative long day (LD) plants at intermediate temperatures (18–24 °C) and obligatory LD plants at high temperature (27 °C). However, at low temperatures (<15 °C), both types are day neutral and flower independently of photoperiod [1,2]. These diverse flowering responses have important implications for how the two plant types should be raised and cultivated.

In Europe, the EB strawberries are mainly used for annual production in plastic tunnels. Usually, runner tips are rooted in late July and raised during autumn under outdoor conditions in specialized plant nurseries in The Netherlands, where the production of such ready-to-flower plants has developed into a considerable industry [3]. Under the relatively low temperature conditions in autumn, the plants will initiate flower buds even in SD. After overwintering in cold storage at −1.5 °C, the plants were established in March–April on a tabletop production system for cropping in plastic tunnels or greenhouses. For convenience, and because of the marginal temperature conditions prevailing in the north, strawberry growers in Nordic countries usually buy their plants from The Netherlands. However, Sønsteby et al. [4] recently reported that plants with the same yield potential as the imported plants could be produced under Nordic temperature conditions according to a slightly modified production protocol.

A serious shortcoming of the European production system is, however, that it does not produce a continuous and stable supply of ripe berries during the harvest season, but rather a series of flowering and fruiting flushes separated by gaps with little or no production [4–7]. The first, and usually largest, fruit flush, which originates from inflorescences produced during plant raising in the previous year [7], is usually followed by an off period of 2–3 or more weeks with little fruit. This causes discontinuous fruit supply and reduced total yields and represents a big challenge in commercial production.

Melis [6] observed that a large first fruit flush produced by large plants with high yield potential was always associated with a long off period, whereas a smaller first flush was combined with a shorter off period. Therefore, he argued that a heavy crop load tended to suppress and delay the initiation of recurrent flower flushes. This was directly supported by the results of Sønsteby et al. [7] who found that both floral initiation and subsequent fruit growth were source-limited in heavily fruiting plants. However, there seemed to be differences in the severity of the problem among cultivars. Thus, the high yielding 'Favori' commonly produces a large first flush, followed by a long off period, whereas the lower yielding 'Murano' has a smaller first fruit flush and a more even distribution of the crop during the fruiting season [7].

This means that the total fruit yield of the EB cultivars is determined by the yield potential established during plant raising, as well as the additional recurrent floral initiation taking place during the cropping season. Accordingly, optimization of fruit yield requires harmonization of the two floral initiation steps (cf. Sønsteby et al. [4]).

Despite the knowledge available on the physiology of flowering of EB strawberries [1], there is need for further investigation on the flower-inducing efficiency of the various temperature and photoperiodic conditions in a range of cultivars under conditions that are relevant for commercial production. This prompted us to perform a controlled environment experiment in which three commercial cultivars were preconditioned at temperatures ranging from 9–27 °C in 10-h and 20-h photoperiods for four weeks during plant raising. Their instant flowering potential, as well as their final yield potential after overwintering in a cold store and cropping in a plastic high tunnel, were assessed.

The aim of the experiment was two-fold: (1) to provide a firm knowledge basis for the flower-inducing efficiency of a range of relevant temperatures and photoperiods in commercial cultivars, and (2) to study the seasonal crop distribution of the treated plants and its relation to total yield.

2. Materials and Methods

2.1. Plant Material and Cultivation

The plant material used for this experiment was propagated in a greenhouse at the NIBIO Experimental Centre Apelsvoll in southeast Norway (60°40′ N, 10°40′ E). Three commercially available and runner-propagated EB strawberry cultivars (*Fragaria* × *ananassa* Duch.) were used for the experiment: Altess (Flevo Berry Holding B.V., Ens, The Netherlands), Favori (Flevo Berry Holding B.V., The Netherlands) and Murano (Conzorcio Italiano Vivaisti, C.I.V., Comacchio, Italy). Young runner plants of all cultivars were collected in mid-July from mother plants grown in a plastic tunnel at NIBIO Apelsvoll. All runner plants were rooted directly in 9 cm pots in a peat-based potting compost (Gartnerjord, LOG, Oslo, Norway) mixed with 20% (*v*/*v*) granulated perlite in a water-saturated atmosphere under a plastic enclosure at 20 h photoperiod and a minimum temperature of 20 °C. On 14 August, all plants were transferred to the phytotron at the Norwegian University of Life Sciences at Ås (59°40′ N, 10°40′ E) and exposed to 10-h short day (SD) and 20-h long day (LD) at temperatures of 9 °C, 15 °C, 21 °C or 27 °C for 4 weeks. In the phytotron, all plants were grown during daytime (8:00 a.m. to 8:00 p.m.) in natural daylight compartments and then moved to adjacent growth rooms from 18:00 h–08:00 h. There, they received either darkness for 14 h (SD) or 10 h low-intensity-light (~7 µmol quanta m^{-2} s^{-1} photosynthetic photon flux (PPF)) from 70 W incandescent lamps for daylight extension (LD), so that the 4 h dark period was centered around midnight (10:00 p.m. to 2:00 a.m.). The daylight extension amounted to less than 2% of the total daily light radiation, all plants thus receiving nearly the same daily light integral in both photoperiods. In the daylight compartments, an additional 125 µmol quanta m^{-2} s^{-1} were automatically added by high-pressure metal halide lamps (400 W Philips HPI-T) whenever the PPF in the compartments fell below 150 µmol quanta m^{-2} s^{-1} (as on cloudy days). The plant trolleys were positioned randomly in the daylight rooms due to the daily movement in and out of the adjacent photoperiodic treatment rooms. Temperatures were controlled to ± 1 °C and a water vapor pressure deficit of 530 Pa was maintained at all temperatures. Throughout the experimental period, the plants were irrigated daily to drip-off with a complete fertilizer solution [electric conductivity 1.2–1.4 mS cm^{-1}, 1:1 YaraTera Kristalon™: YaraLiva Kalksalpeter™ (Yara, Oslo, Norway)].

After this four-week preconditioning treatment, all plants were potted in 12 cm pots with a peat based potting compost, and three plants from each replicate and treatment were forced directly in a greenhouse for 10 weeks with 20 h LD at 20 °C for assessment of their instant flowering status. In the greenhouse, the plants received daylight plus a daily mix of artificial light of approx. 150 µmol quanta m^{-2} s^{-1} PPF from 400 W Philips HPI-T metal halide lamps (8:00 a.m. to 6:00 p.m.) plus about 15 µmol quanta m^{-2} s^{-1} light from 70 W incandescent lamps (6:00 p.m. to 10:00 p.m. and 2:00 a.m. to 8:00 a.m.) throughout the forcing period. The rest of the plants were moved outdoors for continued floral initiation and stabilization from 12 September to 24 October 2018. The daily mean temperatures during this period are shown in Figure 1A. Thereafter, one plant from each replicate was forced for 8 weeks under the same conditions as explained above for assessment of flowering and yield potential. Due to the limited number of 'Murano' plants, this treatment had to be omitted for this cultivar. The remaining plants were cold stored in darkness at -1.5 °C during the period 24 October 2018–5 May 2019.

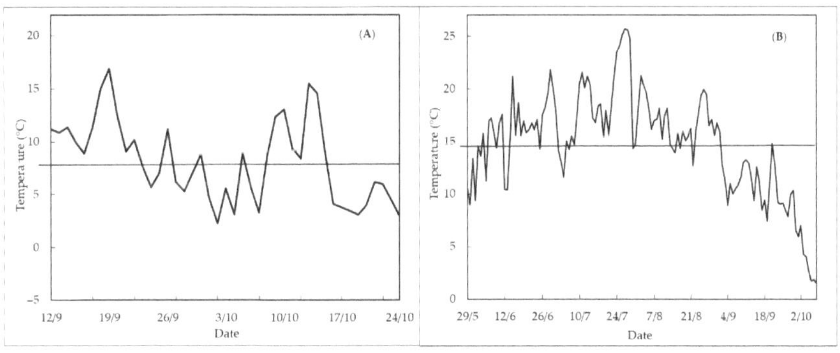

Figure 1. Daily mean temperatures during the outdoor period of plant raising in 2018 (**A**), and in the plastic tunnel during the cropping season in 2019 (**B**). The horizontal lines represent the average daily mean for the respective periods.

The cold-stored plants were then cropped in a tabletop system in a plastic high tunnel for assessment of yield and crop distribution. On 5 May 2019, all plants were transplanted into 2.5 L plastic pots (one plant per pot) in a mixture of 80% limed and fertilized peat and 20% granulated perlite. After a 9-day establishment period in an unheated plastic greenhouse under a double layer of fiber cloth, the plants were placed in an open plastic tunnel on 16 May, and drip irrigated with a nutrient solution containing a 1.1 mixture of YaraTera CALCINIT® and YaraTera Kristalon® (Yara, Norway) with electric conductivity of 1.6 mS cm^{-1}. The daily mean temperatures in the tunnel during the cropping season are shown in Figure 1B.

2.2. Experimental Design and Data Observations

During preconditioning in the phytotron, the experiment was conducted as a randomized block design with three replicates with 10 plants of each cultivar in each treatment. Growth was monitored by weekly registration of the number of leaves, crowns, and runners. At termination of the preconditioning treatments and before forcing or cold storage, the total number of leaves, runners, flower trusses and flowers, and petiole length of the last fully developed leaf were also recorded. During forcing, flowering and growth performance were assessed by weekly recordings of the total number of leaves, runners, flower trusses and open flowers. In the plastic tunnel, the experiment was conducted as a randomized block design with three replicates of 4 plants of each cultivar in each treatment. Ripe berries were harvested 2–3 times per week from 5 July to 2 October. The number and weight of all berries, including unsalable berries, were recorded, as well as the proportion of healthy berries with diameter > 25 mm. In addition, runners were registered and removed throughout the season. At termination of the experiment on 2 October, plant height (measured from base to top of the leaf canopy), number of crowns, leaves per plant and plant fresh weight (excluding runners and roots) were recorded on all plants, as were the number of flowers and berries not reaching maturity.

Experimental data were subjected to analysis of variance (ANOVA) using the MiniTab® v18.1 Statistical Software program package (Minitab Inc., State College, PA, USA). Prior to the analyses, homoscedasticity and normality assumptions were tested (Ryan–Joiner test for normality and Levene's test for homoscedasticity). Percentage values were always subjected to square root transformation before performing the ANOVA.

3. Results

3.1. Flowering and Runnering in the Phytotron

During the preconditioning treatment in the phytotron, open flowers started to appear after three weeks of treatment (after one week at high temperature in 'Murano'), indicating that the plants were induced to flower before the treatments were started (Figure 2). This is typical for EB strawberries, in which flowers usually appear as soon as the runners are formed [2]. The number of flowers was highest in 'Murano' and increased over time with increases in temperature and photoperiod in all cultivars. Runners emerged after two weeks in all cultivars, and the highest number was observed in 'Favori' at intermediate to high temperatures in both photoperiods. The main effects of temperature and cultivar, as well as the two-factor interaction between cultivar and temperature, were all significant (Appendix A, Table A1).

3.2. Flowering Potential of the Preconditioned Plants

Plants from all cultivars and preconditioning treatments started to flower after 2 weeks of forcing, their numbers increasing significantly over time with increases in temperature and photoperiod (Figure 3, Table A2). In all cultivars and at all temperatures, the number of flowers increased in LD, being highest at 15 °C in 'Altess' and 'Murano', and at 21 °C in 'Favori'. Under SD conditions, flowering was strongly delayed at 27 °C in 'Favori' and 'Murano', and to a lesser extent, this was the case at all temperatures in all cultivars. However, after a time span of approximately seven weeks, a new flush of flowers emerged in these plants. Most likely, these flowers were initiated during the forcing treatment in LD at 20 °C. At both forcing times, the main effects of both temperature and photoperiod were significant, with no significant cultivar interactions (Appendix A, Table A2).

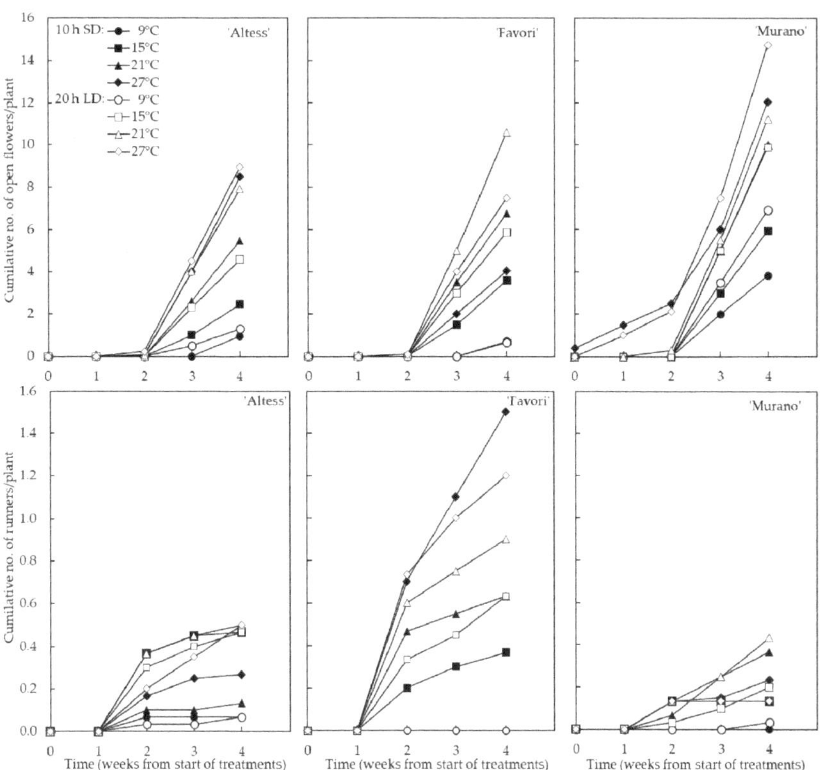

Figure 2. Time courses of cumulative flower and runner appearances during the 4-week preconditioning period in three EB strawberry cultivars. Data are the means of three replicates, each with 10 plants of each cultivar.

Runners started to emerge after two weeks of forcing in all cultivars but remained generally low in plants preconditioned in LD. The number of runners was largest in 'Altess' and 'Murano', in which it increased rapidly during the first 6 weeks of forcing in plants preconditioned in SD at intermediate to high temperatures. Due to varying interactions between temperature and photoperiod, and between the environments and cultivars, the main effects of temperature and photoperiod were not always significant but were generally strongest in the first forcing (Table A2). The results revealed an opposite relationship between flowering and runner formation, with flowering being promoted by LD and intermediate temperatures while runner formation was promoted by SD at the same temperature range.

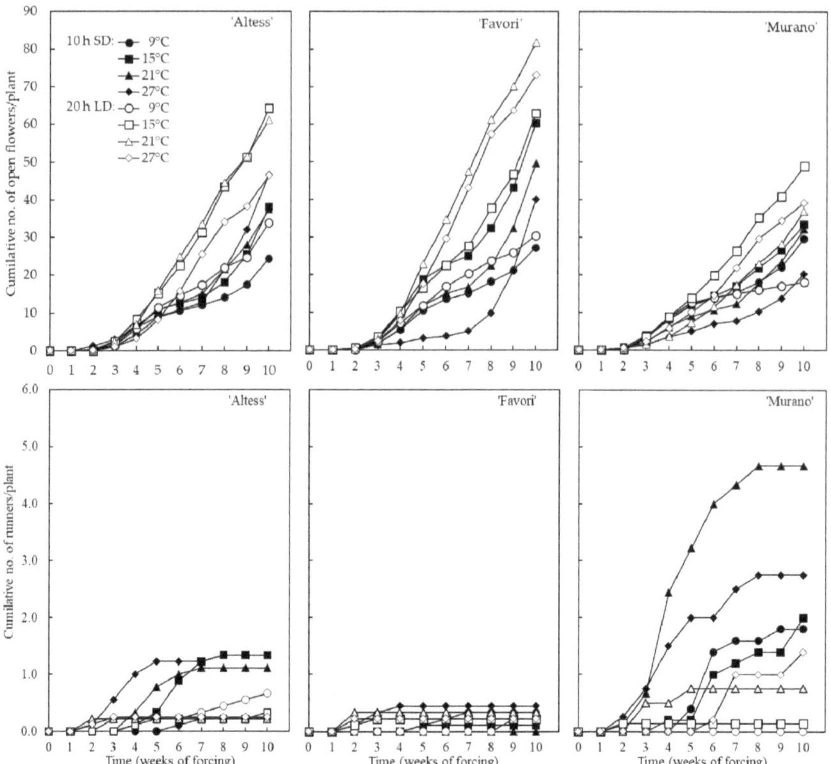

Figure 3. Time courses of cumulative flower (top panel) and runner formation in three EB strawberry cultivars as affected by 4 weeks of temperature and photoperiod preconditioning, as indicated during 10 weeks of subsequent forcing in a greenhouse in 20-h LD at 20 °C. The data are means of three replicates, each with three plants of each cultivar.

3.3. Plant Flowering Potential in Late Autumn

Time courses of flower and runner emergences during an 8-week forcing period started on 24 October (at the end of the raising period) are shown in Figure 4 and Table A2. Due to an insufficient number of 'Murano' plants, only data for 'Altess' and 'Favori' are available. In all plants of both cultivars, flowers started to emerge after 3 weeks of forcing, while the number of flowers varied widely in the various treatments, being highest after preconditioning in LD and intermediate to high temperatures. In SD, on the other hand, flowering was generally sparse and, under all conditions, flowering started to level off after 3–4 weeks. However, after 7 weeks of forcing, a new flush of flowers emerged in all plants regardless of pretreatment conditions. Since the number of additional flowers increased in parallel and started at the same time in all treatments, they were apparently initiated during the flower-inducing (LD and 20 °C) forcing treatment (cf. Figure 3). Generally, the two cultivars responded in the same way; only the main effects of temperature and photoperiod were statistically significant (Table A2).

Runners started to emerge shortly after the forcing was started in both cultivars but remained low for the first 4–5 weeks (Figure 4). From week 5 onwards, the number of runners increased somewhat again in the sparsely flowering plants preconditioned in SD. The main effect of temperature and the two-factor interaction between temperature and photoperiod were statistically significant (Table A2).

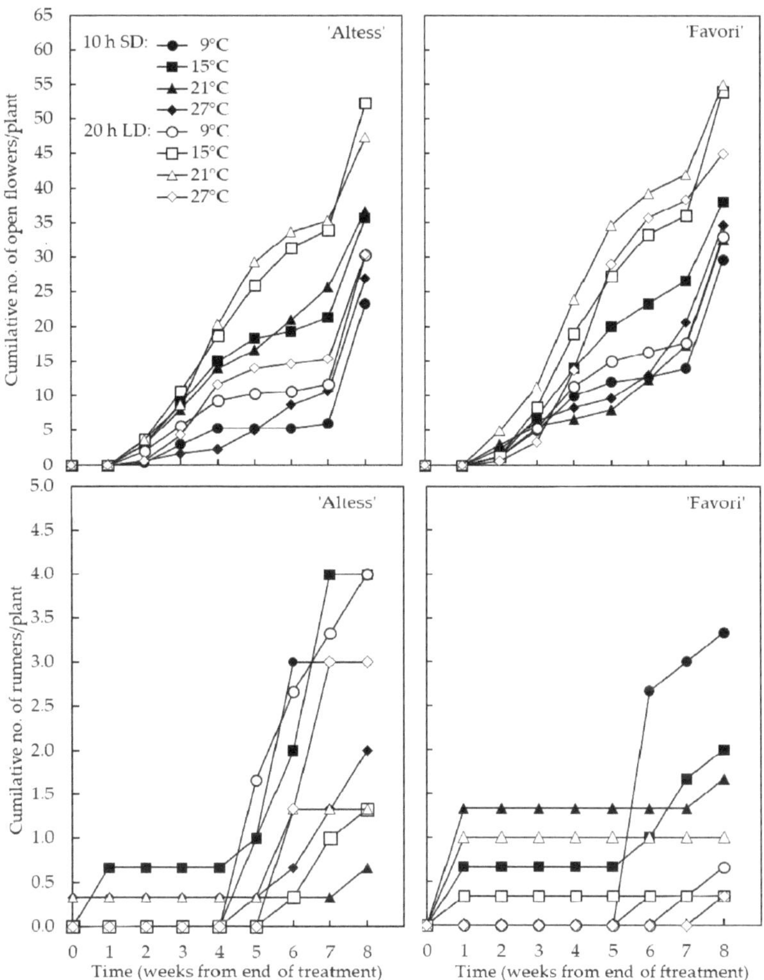

Figure 4. Time courses of cumulative flower (top panel) and runner formation in two EB strawberry cultivars as affected by 4 weeks of temperature and photoperiod pre-conditioning as indicated, followed by further raising under natural outdoor conditions from 12 September until 24 October. The plants were forced for 8 weeks in a greenhouse in 20-h LD at 20 °C. The data are means of three replicates, each with one plant of each cultivar.

3.4. Yield Performance of the Preconditioned Plants

The highest berry yield of 1.592 g/plant was recorded in 'Favori' plants preconditioned in LD at 15 °C. 'Favori' also had the highest total berry yield across all temperatures and daylengths, whereas 'Murano' had the lowest total yields (Table 1). In general, the yields were highest at intermediate temperatures (15–21 °C), and lowest at 27 °C, while the effect of photoperiod varied with cultivar and temperature. For 'Altess', the highest total yields were obtain in plants preconditioned in SD across the range of temperatures. For 'Favori', the highest total yields were obtained in plants preconditioned in LD at temperatures up to 21 °C, while at 27 °C, the yields were higher in SD. For 'Murano', the highest total yields were obtained in LD at intermediate temperatures, while in SD at temperatures of 9 °C and 27 °C.

Table 1. Effects of preconditioning temperature and photoperiod on total berry yield and yield components of three EB strawberry cultivars when cropped on a tabletop system in high plastic tunnels in the year after treatment. Data are means ± SD of three replications, each with four plants of each cultivar. Plants were harvested from 8 July to 2 October 2019.

Cultivar	Temp. (°C)	Yield (g/Plant)		Berry Weight (g)		Berries Plant^{-1}	
		Photoperiod (h)					
		10	20	10	20	10	20
Altess	9	1276.5 ± 147.4	1235.6 ± 82.6	20.2 ± 0.6	18.2 ± 1.0	63.3 ± 7.5	67.9 ± 5.1
	15	1266.2 ± 211.2	1261.3 ± 53.9	18.2 ± 1.3	16.8 ± 0.6	70.2 ± 16.2	75.1 ± 5.1
	21	1204.9 ± 98.1	1143.7 ± 331.7	18.6 ± 0.4	15.9 ± 0.7	64.8 ± 5.3	71.8 ± 20.1
	27	1134.7 ± 204.8	1054.7 ± 352.1	17.6 ± 0.7	15.6 ± 0.4	64.4 ± 12.3	67.6 ± 22.9
	Mean	1220.6	1173.8	18.7	16.6	65.7	70.6
Favori	9	1318.5 ± 12.9	1320.1 ± 20.5	18.0 ± 1.2	18.3 ± 0.3	73.3 ± 4.6	72.3 ± 0.8
	15	1193.9 ± 99.3	1592.4 ± 203.0	18.1 ± 0.7	17.0 ± 1.1	66.0 ± 7.9	93.4 ± 6.6
	21	1216.6 ± 124.7	1280.6 ± 263.8	17.8 ± 0.9	15.6 ± 0.3	68.6 ± 8.8	82.3 ± 17.8
	27	1149.6 ± 234.1	1105.1 ± 439.2	17.1 ± 0.9	13.5 ± 0.3	67.8 ± 17.4	81.5 ± 32.3
	Mean	1219.6	1324.5	17.8	16.1	68.9	82.4
Murano	9	1063.0 ± 177.3	864.5 ± 634.2	19.0 ± 1.9	14.9 ± 3.2	56.4 ± 12.7	54.0 ± 30.8
	15	1065.0 ± 296.6	1219.1 ± 107.4	19.0 ± 0.4	17.6 ± 0.6	56.1 ± 16.9	69.2 ± 5.9
	21	848.2 ± 247.7	1114.1 ± 192.5	17.0 ± 1.3	17.4 ± 1.0	50.4 ± 16.3	64.0 ± 8.2
	27	1065.7 ± 180.5	589.5 ± 116.4	18.7 ± 0.6	14.4 ± 1.2	57.1 ± 11.0	40.8 ± 6.9
	Mean	1010.0	946.8	18.5	16.1	55.0	57.0
Probability level of significance (ANOVA)							
Source of variation							
Temperature (A)		<0.001		ns		ns	
Photoperiod (B)		<0.001		<0.001		<0.001	
A × B		0.03		<0.001		ns	
Cultivar (C)		<0.001		<0.001		0.001	
C × A		ns		ns		0.001	
C × B		ns		ns		ns	
A × B × C		ns		ns		0.04	

Table 1. *Cont.*

Cultivar	Temperature (°C)	Berries >25 mm (%)		Unsaleable Berries (g/plant)	
		Photoperiod (h)			
		10	20	10	20
Altess	9	99.6 ± 0.2	98.6 ± 1.0	19.4 ± 2.9	32.4 ± 3.6
	15	99.0 ± 0.4	98.6 ± 0.4	19.3 ± 3.7	20.8 ± 4.8
	21	99.2 ± 0.2	98.3 ± 0.6	12.3 ± 4.8	18.0 ± 4.9
	27	98.1 ± 0.5	98.0 ± 0.4	11.6 ± 3.7	14.1 ± 5.7
	Mean	99.0	98.4	15.6	21.3
Favori	9	99.0 ± 0.4	98.6 ± 0.6	17.1 ± 3.5	8.8 ± 2.0
	15	99.0 ± 0.9	98.1 ± 0.8	14.3 ± 3.9	9.4 ± 2.9
	21	98.2 ± 0.9	97.0 ± 0.4	25.4 ± 8.6	8.4 ± 2.7
	27	97.6 ± 0.3	94.8 ± 0.8	16.8 ± 2.1	22.4 ± 3.1
	Mean	98.5	97.1	18.4	12.3
Murano	9	98.2 ± 0.6	94.4 ± 5.8	28.2 ± 9.2	21.0 ± 2.7
	15	98.8 ± 0.3	97.2 ± 0.9	25.8 ± 11.1	19.9 ± 9.9
	21	96.6 ± 1.4	96.5 ± 0.7	8.9 ± 3.1	15.0 ± 2.3
	27	97.1 ± 1.8	95.4 ± 0.8	9.9 ± 5.7	2.3 ± 1.1
	Mean	97.7	95.9	18.2	12.6
Probability level of significance (ANOVA)					
Source of variation					
Temperature (A)		ns		ns	
Photoperiod (B)		0.005		0.003	
A × B		0.01		ns	
Cultivar (C)		<0.001		0.007	
C × A		ns		ns	
C × B		ns		ns	
A × B × C		ns		ns	

ns, not significant.

The yield components, berry weight (size) and number of berries per plant, varied in an inverse relationship in all cultivars, the former being generally enhanced by SD and the latter by LD preconditioning at all temperatures (some variation in 'Murano', however) (Table 1). Berry weight decreased with increasing temperature in both photoperiods, but due to a highly significant interaction of photoperiod and temperature, the main effect of temperature was not statistically significant. The number of berries per plant varied significantly between the cultivars, being lower in 'Murano' than in the other two cultivars (highly significant interaction between cultivar and temperature).

As shown in Figure 5, the temporal distribution of the berry production and the size of the fruiting flushes differed widely between cultivars and treatments. There was a clear relationship between the size of the first fruit flush and the size and temporal distribution of the rest of the crop. A large and concentrated first fruit flush was always associated with a marked subsequent off period with little or no fruit. In 'Altess' and particularly in 'Favori', the severity and duration of the off period increased markedly with increasing preconditioning temperature. In fact, 'Favori' plants preconditioned in LD at 27 °C, recurrent flowering and fruiting never fully recovered during the season. In 'Murano', which generally had relatively small first fruit flushes and a more even distribution of the crop during the harvest season, this effect was only present in plants preconditioned at 27 °C in SD. Generally, smaller first fruiting flushes led to more stable berry yields during the rest of the season. For all three cultivars, the first ripe fruits appeared after 5 weeks of cultivation in the polytunnel. Plants preconditioned at 9 °C always had a small first flush regardless of photoperiod in all cultivars, but had relatively constant berry production during the cropping season (Figure 5).

The time courses of runner production during the cropping season shown in Figure 6, demonstrating a general declining seasonal trend in all cultivars. A large share of the runners were produced before week 28 when the berry harvest started, and the majority appeared during the first half of the season. 'Favori' produced less runners than the other two cultivars, and in all cultivars the highest number of runners was produced in plants preconditioned in SD (Table 2). At intermediate to high temperatures, runners and flowers were produced in parallel in all cultivars during weeks 28 to 31, while at 9 °C, there was an opposite trend between runner and flower production in 'Murano' and 'Altess'. For 'Favori', there was a decreasing trend for both runner and flower production during the same period.

Some measures of plant architecture at the end of the harvest season are presented in Table 2. All the parameters measured varied significantly between the cultivars. The growth-related parameters (plant height, crowns, leaves and runners per plant), as well as plant fresh weight and inflorescences per plant, were always enhanced by SD and were highest in 'Favori'. The main effect of temperature was significant only for plant height, inflorescences per plant and flowers/fruits not harvested. In all cultivars, the total number of inflorescences were usually highest at intermediate temperatures (21 °C and 15 °C) in both daylengths. The number of flowers and fruits that did not reach maturity before the harvest was terminated on 8 October was significantly higher in plants preconditioned in SD at intermediate temperatures (Table 2).

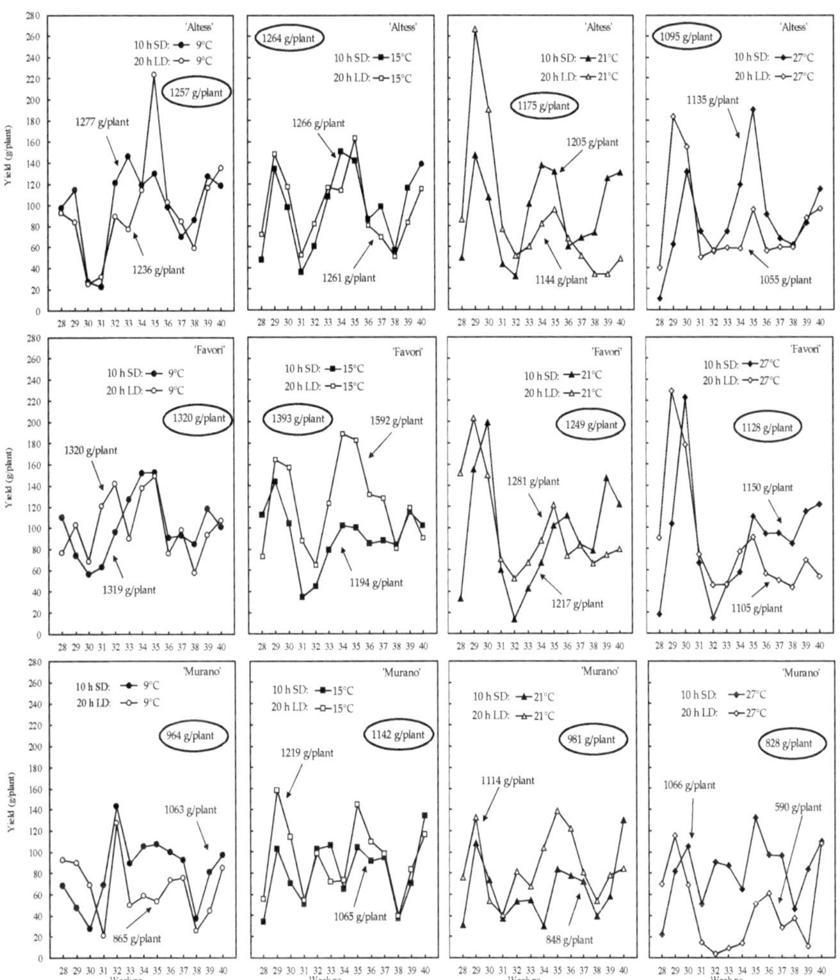

Figure 5. Time courses of weekly berry yield of three EB strawberry cultivars as affected by 4 weeks of preconditioning at temperatures of 9 °C, 15 °C, 21 °C, 27 °C and 10 h or 20 h photoperiods, as indicated, followed by further raising under natural outdoor conditions from 12 September until 24 October 2018. The plants were cropped in a plastic tunnel after overwintering in cold storage at −1.5 °C from 24 October 2018 until 5 May 2019. The data are means of three replications with four plants each. Values in the ovals represent the mean yield for both photoperiods at each temperature.

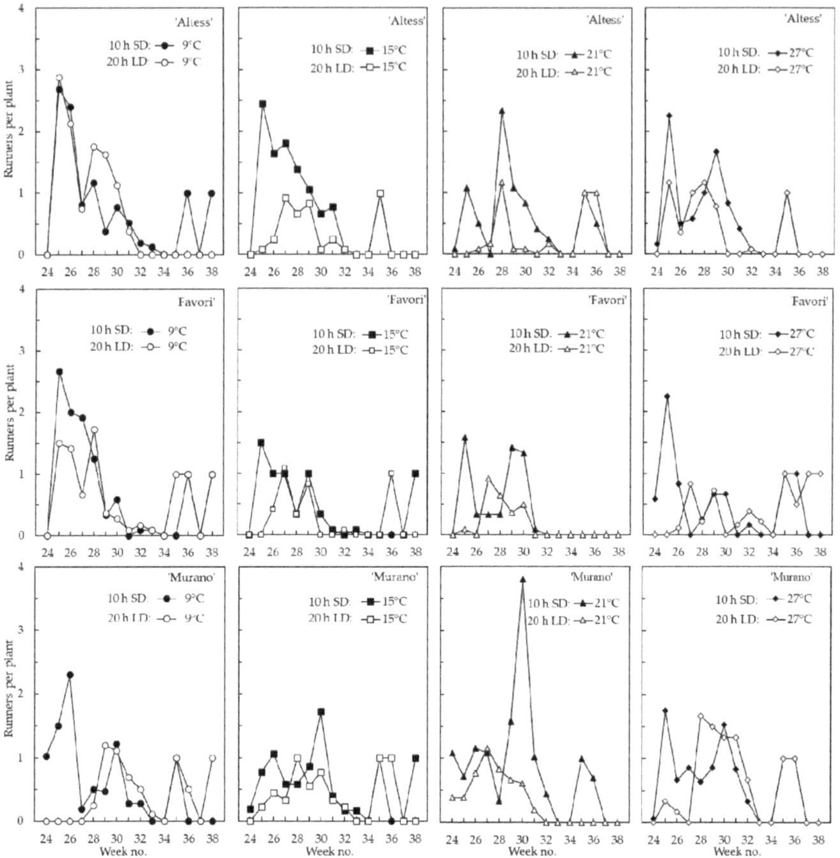

Figure 6. Time courses of weekly runner production of three EB strawberry cultivar as affected by 4 weeks of preconditioning at temperatures of 9 °C, 15 °C, 21 °C, 27 °C and 10 h or 20 h photoperiods, followed by further raising under natural outdoor conditions from 12 September until 24 October 2018. The plants were cropped in a plastic tunnel after overwintering in cold storage at −1.5 °C from 24 October 2018 until 5 May 2019. The data are means of three replications with four plants each.

Table 2. Effects of raising temperature and photoperiod on plant architecture in three EB strawberry cultivars grown on a tabletop system in a plastic high tunnel in the year after treatment. Data are means ± SD of three replicates, each with four plants of each cultivar.

Cultivar	Temp. (°C)	Plant Height (cm)		Crowns Plant^{-1}		Leaves Plant^{-1}		Runners Plant^{-1} *	
		Photoperiod (h)							
		10	20	10	20	10	20	10	20
Altess	9	28.0 ± 2.3	28.5 ± 0.7	4.0 ± 1.0	4.1 ± 0.8	31.4 ± 6.3	37.0 ± 2.8	5.6 ± 1.3	7.8 ± 1.7
	15	28.0 ± 1.4	25.9 ± 2.8	4.5 ± 1.2	4.4 ± 0.8	32.8 ± 8.8	35.8 ± 0.5	8.4 ± 1.5	2.9 ± 0.4
	21	27.0 ± 1.8	23.9 ± 2.1	3.9 ± 0.1	3.0 ± 0.6	35.4 ± 4.2	24.4 ± 6.2	7.1 ± 1.8	2.1 ± 0.7
	27	27.1 ± 4.1	23.1 ± 3.7	4.6 ± 0.1	2.9 ± 1.2	32.5 ± 1.4	26.7 ± 8.1	6.6 ± 1.3	4.4 ± 1.8
	Mean	27.5	25.4	4.3	3.6	33.2	30.9	7.0	4.3
Favori	9	29.8 ± 3.4	29.8 ± 0.6	5.9 ± 0.8	3.8 ± 0.4	38.9 ± 7.2	35.8 ± 1.9	6.9 ± 1.8	4.6 ± 1.9
	15	33.2 ± 2.7	33.2 ± 3.2	5.3 ± 1.8	6.3 ± 0.7	41.6 ± 10.6	51.8 ± 6.6	4.3 ± 2.2	2.3 ± 1.7
	21	32.8 ± 3.3	26.0 ± 1.1	5.5 ± 0.3	5.1 ± 0.5	57.8 ± 2.6	41.7 ± 6.4	5.1 ± 0.4	2.5 ± 1.1
	27	31.3 ± 3.6	22.9 ± 2.9	7.3 ± 0.8	4.3 ± 1.7	55.3 ± 7.7	31.2 ± 9.3	4.6 ± 1.5	2.6 ± 1.3
	Mean	31.8	28.0	6.0	4.8	48.4	40.1	5.2	3.0
Murano	9	23.1 ± 2.3	19.6 ± 3.8	4.8 ± 1.4	4.2 ± 2.0	30.6 ± 4.5	35.4 ± 17.4	5.5 ± 0.9	5.1 ± 1.1
	15	24.4 ± 2.5	23.6 ± 3.2	4.8 ± 1.6	4.0 ± 1.3	37.3 ± 8.6	28.3 ± 8.5	5.4 ± 0.5	3.4 ± 1.1
	21	26.9 ± 2.7	21.8 ± 1.9	6.3 ± 1.4	3.6 ± 0.5	44.1 ± 9.4	26.1 ± 7.1	10.1 ± 3.5	4.3 ± 1.5
	27	23.6 ± 2.5	17.2 ± 5.8	7.3 ± 1.4	5.2 ± 2.0	43.9 ± 19.3	33.3 ± 17.4	6.9 ± 2.3	6.8 ± 3.6
	Mean	24.5	20.5	5.8	4.2	39.0	30.8	7.0	4.9

Probability level of significance (ANOVA)
Source of variation

Source of variation	Plant Height (cm)	Crowns Plant^{-1}	Leaves Plant^{-1}	Runners Plant^{-1} *
Temperature (A)	<0.010	ns	ns	ns
Photoperiod (B)	<0.001	0.042	0.038	<0.001
A × B	0.001	ns	0.006	0.019
Cultivar(C)	0.022	0.044	0.010	0.008
C × A	ns	ns	ns	ns
C × B	ns	ns	ns	ns
A × B × C	ns	ns	ns	ns

Table 2. *Cont.*

Cultivar	Temp. (°C)	Plant FW (g)		Infloresc. Plant^{-1} *		Flowers/Fruits not Harvested	
		Photoperiod (h)					
		10	20	10	20	10	20
Altess	9	287.1 ± 21.1	295.2 ± 54.0	12.2 ± 0.1	12.2 ± 2.6	70.7 ± 11.3	78.3 ± 14.5
	15	267.6 ± 30.0	211.3 ± 5.5	12.8 ± 1.6	14.5 ± 0.7	85.1 ± 14.3	62.9 ± 6.0
	21	277.0 ± 23.9	146.6 ± 43.1	14.3 ± 0.6	11.3 ± 2.3	81.8 ± 14.6	61.3 ± 13.9
	27	261.2 ± 15.6	158.1 ± 56.7	13.4 ± 0.8	11.3 ± 1.7	74.3 ± 12.3	61.5 ± 17.0
	Mean	271.9	202.8	13.3	12.3	78.6	66.0
Favori	9	297.3 ± 61.4	262.3 ± 16.3	13.6 ± 3.1	11.3 ± 0.7	63.5 ± 20.2	52.8 ± 13.7
	15	320.6 ± 66.3	341.9 ± 62.2	16.3 ± 3.4	17.4 ± 2.4	88.6 ± 43.8	71.8 ± 26.5
	21	411.9 ± 34.8	231.6 ± 35.8	19.4 ± 1.3	18.6 ± 4.5	117.7 ± 6.5	76.6 ± 10.4
	27	385.2 ± 56.0	179.7 ± 42.2	18.3 ± 3.4	13.7 ± 1.5	112.3 ± 31.7	66.1 ± 5.5
	Mean	353.8	253.9	16.9	15.2	95.5	66.8
Murano	9	185.0 ± 14.8	157.3 ± 94.2	12.6 ± 1.9	9.5 ± 3.5	76.3 ± 17.0	66.0 ± 24.1
	15	228.6 ± 41.5	157.8 ± 57.5	14.6 ± 3.9	11.8 ± 1.7	114.2 ± 15.9	67.7 ± 14.9
	21	242.8 ± 46.6	161.9 ± 35.4	17.6 ± 1.7	12.7 ± 2.7	128.2 ± 18.1	59.3 ± 10.4
	27	211.3 ± 30.1	163.5 ± 68.3	14.6 ± 6.2	15.2 ± 6.3	98.7 ± 41.0	112.5 ± 53.5
	Mean	216.9	160.1	14.8	12.3	104.3	76.4
Probability level of significance (ANOVA)							
Source of variation							
Temperature (A)		ns		0.019		0.042	
Photoperiod (B)		<0.001		ns		0.022	
A × B		0.002		ns		ns	
Cultivar(C)		<0.001		0.044		ns	
C × A		ns		ns		ns	
C × B		ns		ns		ns	
A × B × C		ns		ns		ns	

ns, not significant. * Total number of runners and inflorescences produced during the cropping season.

4. Discussion

4.1. Flowering and Runnering in the Phytotron

The number of runners and flowers produced during the 4-week preconditioning period (Figure 2) was merely attributed to the previous environmental history of the stock plants from which the runners were taken. Accordingly, their time of emergence was enhanced by increasing temperatures during the treatment period. It is well documented that, due to their perpetual flowering nature, the runnering capacity of EB strawberry cultivars is generally low [2,7]. Although Rivero et al. [2] and Sønsteby et al. [4] showed that 'Favori' produced relatively few runners, the cultivar had more runners than 'Altess' and 'Murano' in the present experiment. Obviously, these deviations were due to the different environmental prehistory of the stock plants. As generally found in EB strawberry cultivars [2,5], flowering and runnering varied in an inverse manner and a rapid increase in flower emergence took place in all cultivars as runner formation levelled off (Figure 2).

4.2. Flowering Potential of the Preconditioned Plants

An inverse relationship between flowering and runnering was demonstrated in all cultivars and most clearly in 'Favori', where a rapid increase in the number of flowers after week 3 of forcing was associated with cessation of runner production in all treatments (Figure 3). A similar situation was revealed in the LD-treated plants of the other cultivars, although a few runners emerged in the sparsely flowering SD-treated plants. In all cultivars, flowering was enhanced by LD and increasing temperatures (except for 'Murano' at 15 °C), whereas SD strongly delayed flowering at high temperatures, especially in 'Favori' and 'Murano'. On the other hand, runner formation was enhanced by SD over the same intermediate to high temperature range. All these results concur with previous results for EB cultivars in general [8,9] as well as for 'Favori' and other modern EB cultivars [2,4,10,11]. Furthermore, a new flush of flowers emerged after a time span of approximately seven weeks, which had apparently been initiated after the plants had been transferred to the LD and relatively high temperature forcing conditions. This second flush of flowering was less pronounced in 'Murano' than in 'Favori' and 'Altess' (Figure 3).

4.3. Plant Flowering Potential in Late Autumn

The flowering and runnering performance of the plants after completion of the raising season was in many ways comparable with those of plants forced immediately after completion of the 4-week preconditioning treatment (cf. Figures 3 and 4). The flower-promoting effects of LD and high temperature preconditioning were still pronounced, and the inverse relationship between flowering and runnering also remained much the same. However, both flowers and runners started to emerge about one week earlier in the later forced plants, and in all treatments, the emergence of flowers levelled off after approximately six weeks of forcing. This was associated with the emergence of a few runners in plants with sparse flowering. Furthermore, as in the directly forced plants, a new flush of flowers emerged after seven weeks of forcing. Since the increase in number of flowers was identical in all preconditioned plants groups, these flowers were clearly initiated during the forcing treatment, as previously suggested for the directly forced plants.

These results demonstrated that recurrent floral initiation readily takes place when previously induced plants with advanced flower primordia are transferred back to inductive conditions. The critical point seemed to be whether the plants had become dormant or not and whether they were exposed to dormancy-breaking chilling. In the present experiment, the plants were exposed to daily mean temperatures ranging from 5–15 °C for six weeks, and the levelling in the number of open flowers (in Figure 4) indicated that plants from all treatments had been at the verge of becoming dormant when the forcing started. This concurred with the results of Rivero et al. (2021) [2] who found that low temperature (6 °C) did not induce dormancy in 'Favori' plants regardless of daylength conditions, while they became dormant after 10 weeks (but not 5 weeks) of SD exposure at 16 and 26 °C. It is also clear that the present temperature conditions did not bring about the delay of recurrent floral initiation that results from long-term chilling at sub-zero temperatures (cf. Gallace et al. [5]).

Another interesting finding was that the number of flowers emerging did not continue to increase when the pretreated plants were grown further under natural outdoor conditions for another six weeks before forcing. In fact, the number of emerging flowers and runners decreased somewhat, even when compensating for the 2-week shorter forcing period (cf. Figures 3 and 4). This indicated that some flower primordia aborted during autumn and suggested a limit to how many viable flowers the plants could accumulate and support until flowering and fruiting. This was contrary to the stabilization of flowering by low temperature that we expected.

4.4. Yield Performance of the Preconditioned Plants

A rather short exposure to varying temperature and photoperiod conditions during the raising of the plants in one year had a remarkable effect on the flowering and fruiting pattern in the following year. The highest yields were generally obtained in plants exposed to LD at intermediate temperatures of 15–21 °C. This temperature range has previously been reported as highly effective for LD-induced flower bud formation in EB cultivars [2,8–10] and was also found to be optimal for photosynthesis in these cultivars [12]. As previously reported by Melis [6], treatments that produced a large first fruiting flush were generally associated with a long off period with little flowering and fruiting. This was particularly pronounced in plants preconditioned in LD at 21 and 27 °C, conditions that are optimal for flower initiation in EB strawberries, whereas treatments that yielded smaller first fruit flushes gave a more balanced and evenly distributed harvest during the season (Figure 5). It is interesting to note that the time lapse between fruiting flushes was usually six to seven weeks, the same time as found for recurrent flowering in greenhouse-forced plants.

As flowers and developing fruits are strong sinks for photosynthates, Sønsteby et al. [4,7] found that a heavy fruit load not only constrained the growth of developing fruits, but also repressed and delayed recurrent initiation of successive flushes, as demonstrated by Melis [6]. While this trend was pronounced in 'Favori' and 'Altess' it was less marked in 'Murano', as also found by Sønsteby et al. [4]. Thus, the present results concurred with previous reports, confirming that the high-yielding 'Favori' was more vulnerable to this phenomenon than the lower yielding 'Murano'. Because of the long off periods associated with large first flushes in plants preconditioned at high temperature, the total fruit yield over the entire harvest season was generally highest in plants preconditioned at 15 °C (at 21 °C in 'Murano'), while it always declined at higher temperatures. As discussed by Sønsteby et al. [7], continuous and excessive flower initiation in LD at high temperatures also resulted in reduced leaf canopy and plant weight. Similarly, Rivero et al. [2] found that 'Favori' was susceptible to overproduction of flowers at optimal flowering conditions, and that this could have detrimental consequences for maintenance of the leaf canopy and perpetual flower initiation. Therefore, as discussed by Sønsteby et al. [4], the magnitude of the first crop flush may have to be compromised in order to optimize total yield and the seasonal distribution of the crop.

The number of runners produced during the cropping season was relatively low, although it increased significantly after winter chilling, as reported by Gallace et al. [5]. The number was lowest in 'Favori', and in all cultivars, a large share was produced during the first five weeks before the fruit harvest started (Figure 6).

5. Conclusions

The results of the present study demonstrated that a brief exposure of 4 weeks to varying temperature and photoperiod conditions during plant raising had a strong effect on the instant flowering potential of the plants and a remarkable long-term effect on the yield and temporal distribution of the harvest in the following year in the EB strawberry cultivars Altess, Favori and Murano. The responses were determined by a pronounced interaction of temperature and photoperiod. As previously reported for these and other

EB cultivars, the perpetual flowering proved to be source-limited, since large first fruiting flushes repressed and delayed successive flushes and strongly reduced plant leaf canopy. Accordingly, the size of the first fruiting flush must be compromised in order to optimize total yield and the seasonal distribution of the harvest. Through proper manipulation of the light and temperature environment during plant raising, it seems feasible to design tailor-made plant production programs for high and continuous yields.

Author Contributions: Conceptualization and methodology, R.R., S.F.R., O.M.H. and A.S.; data curation software, A.S., O.M.H. and R.R.; validation, R.R., S.F.R., O.M.H. and A.S.; formal analysis, A.S. and R.R.; investigation, R.R.; resources, A.S. and S.F.R.; writing—original draft preparation, R.R.; writing—review and editing, R.R., S.F.R., O.M.H. and A.S.; visualization, R.R., S.F.R., O.M.H. and A.S.; supervision, R.R., S.F.R., O.M.H. and A.S.; project administration, A.S. and S.F.R.; funding acquisition, A.S. and S.F.R. All authors have read and agreed to the published version of the manuscript.

Funding: This research was funded by the PhD programme at the Norwegian University of Life Sciences, and the Norwegian Agricultural Agreement Research Fund/Foundation for Research Levy on Agricultural Products, grant number 280608 and Grofondet, grant number 170008.

Institutional Review Board Statement: Not Applicable.

Informed Consent Statement: Not Applicable.

Data Availability Statement: The plant and yield data are available upon request from the corresponding author.

Acknowledgments: We thank Unni M. Roos, Signe Hansen and Kari Grønnerød for excellent technical assistance.

Conflicts of Interest: The authors declare no conflict of interest.

Appendix A

Table A1. Appearances of flowers and runners in three EB strawberry cultivars during the 4-week preconditioning treatment in the phytotron. The data are means $\pm$ SD of three replicates, each with 10 plants of each cultivar.

Cultivar	Temperature (°C)	Flowers Plant^{-1}		Runners Plant^{-1}	
		Photoperiod (h)			
		10	20	10	20
Altess	9	1.0 ± 0.4	1.3 ± 0.4	0.1 ± 0.0	0.1 ± 0.0
	15	2.5 ± 0.8	4.6 ± 0.4	0.5 ± 0.2	0.5 ± 0.1
	21	5.5 ± 2.4	8.0 ± 1.0	0.1 ± 0.0	0.5 ± 0.2
	27	8.6 ± 1.3	8.9 ± 2.8	0.3 ± 0.1	0.5 ± 0.1
	Mean	4.4	5.7	0.2	0.4
Favori	9	0.7 ± 0.1	0.7 ± 0.1	$0.0 \pm$	0.0 ± 0.0
	15	3.6 ± 0.0	5.9 ± 1.2	$0.4 \pm$	0.6 ± 0.1
	21	6.8 ± 0.9	10.6 ± 1.4	$0.6 \pm$	0.9 ± 0.3
	27	4.0 ± 0.9	7.6 ± 1.8	$1.5 \pm$	1.2 ± 0.2
	Mean	3.8	6.2	0.6	0.7
Murano	9	3.8 ± 0.9	7.1 ± 1.5	0.0 ± 0.0	0.0 ± 0.0
	15	5.9 ± 1.7	10.0 ± 2.5	0.1 ± 0.0	0.2 ± 0.0
	21	10.0 ± 3.2	11.7 ± 1.3	0.4 ± 0.0	0.4 ± 0.1
	27	13.6 ± 5.4	16.9 ± 1.4	0.2 ± 0.0	0.1 ± 0.0
	Mean	8.3	11.4	0.2	0.2
Probability level of significance (ANOVA)					
Source of variation					
Temperature (A)		<0.001		<0.001	
Photoperiod (B)		0.003		ns	
A $\times$ B		ns		ns	
Cultivar (C)		<0.001		<0.001	
C $\times$ A		0.002		<0.001	
C $\times$ B		ns		ns	
A $\times$ B $\times$ C		ns		ns	

ns, not significant.

Table A2. Effects of temperature and photoperiod during plant raising on growth and flowering of three EB strawberry cultivars as assessed by 10 weeks of forcing started immediately after 4 weeks of preconditioning, and by 8 weeks of forcing started in late autumn after completion of plant raising (preconditioning + outdoor treatment). In both cases the plants were forced in a heated greenhouse maintained at 20 h light and 20 °C. The data are means ±SD of three replicates, each with 3 or 1 plant respectively, of each cultivar.

		After Preconditioning				In Late Autumn			
		Flowers Plant^{-1}		Runners Plant^{-1}		Flowers Plant^{-1}		Runners Plant^{-1}	
Cultivar	Temperature (°C)	Photoperiod (h)							
		10	20	10	20	10	20	10	20
Altess	9	42.9 ± 6.9	56.9 ± 13.0	0.3 ± 0.0	0.7 ± 0.3	23.3 ± 6.0	30.3 ± 4.0	3.0 ± 1.0	4.0 ± 1.0
	15	63.0 ± 8.3	87.8 ± 23.0	1.3 ± 0.2	0.5 ± 0.0	35.7 ± 9.1	52.3 ± 14.0	4.0 ± 0.0	2.0 ± 0.0
	21	58.8 ± 1.9	83.3 ± 23.0	1.1 ± 0.6	0.7 ± 0.0	36.7 ± 7.2	47.3 ± 8.5	1.0 ± 0.0	2.0 ± 0.4
	27	75.2 ± 14.5	67.3 ± 21.2	1.3 ± 0.6	0.4 ± 0.1	27.0 ± 3.5	30.3 ± 7.2	2.0 ± 0.8	3.0 ± 1.7
	Mean	60.0	73.8	1.1	0.6	30.7	40.1	2.6	2.9
Favori	9	49.9 ± 8.2	54.9 ± 6.9	0.5 ± 0.1	0.3 ± 0.0	29.7 ± 2.3	33.0 ± 7.2	3.3 ± 1.1	2.0 ± 0.0
	15	91.1 ± 3.7	99.9 ± 14.6	0.3 ± 0.0	0.3 ± 0.0	38.0 ± 7.0	54.0 ± 4.4	2.0 ± 0.1	1.0 ± 0.0
	21	81.7 ± 5.5	109.2 ± 27.1	0.0 ± 0.0	0.5 ± 0.0	32.7 ± 4.9	55.0 ± 7.9	1.7 ± 0.2	1.5 ± 0.7
	27	79.8 ± 16.6	92.8 ± 5.4	0.4 ± 0.1	0.7 ± 0.0	34.7 ± 7.2	45.0 ± 7.0	1.0 ± 0.0	1.0 ± 0.0
	Mean	75.6	89.2	0.4	0.4	33.8	46.8	2.2	1.4
Murano	9	43.5 ± 13.7	19.0 ± 8.5	2.3 ± 1.0	0.0 ± 0.0	-	-	-	-
	15	47.6 ± 1.9	59.7 ± 22.7	2.7 ± 0.5	0.3 ± 0.0	-	-	-	-
	21	47.0 ± 13.1	52.7 ± 18.3	4.7 ± 2.4	3.0 ± 0.0	-		-	-
	27	33.8 ± 6.8	44.8 ± 21.9	3.8 ± 1.1	1.3 ± 0.5	-	-	-	-
	Mean	43.8	46.3	3.5	1.5	-	-	-	-
Probability level of significance (ANOVA)									
Source of variation									
Temperature (A)		0.008		ns		0.004		0.041	
Photoperiod (B)		0.03		<0.001		0.001		ns	
A × B		ns		ns		ns		0.030	
Cultivar (C)		<0.001		<0.001		ns		ns	
C × A		ns		0.04		ns		ns	
C × B		ns		<0.001		ns		ns	
A × B × C		ns		ns		ns		ns	

ns, not significant.

References

1. Heide, O.M.; Stavang, J.A.; Sønsteby, A. Physiology and genetics of flowering in cultivated and wild strawberries-a review. *J. Hortic. Sci. Biotec.* **2013**, *88*, 1–18. [CrossRef]
2. Rivero, R.; Remberg, S.F.; Heide, O.M.; Sønsteby, A. Environmental Regulation of Dormancy, Flowering and Runnering in Two Genetically Distant Everbearing Strawberry Cultivars. *Sci. Hortic.* **2021**, *290*, 110515. [CrossRef]
3. Lieten, P. The strawberry nursery industry in The Netherlands: An update. *Acta Hortic.* **2014**, *1049*, 99–106. [CrossRef]
4. Sønsteby, A.; Sadojevic, M.; Heide, O.M. Effect of rooting date and environmental conditions during plant raising on fruit yield performance of two everbearing strawberry cultivars in a Nordic climate. *Horticulturae* **2022**, *8*, 249. [CrossRef]
5. Gallace, N.; Lieten, P.; Boonen, M.; Bylemans, D. Reduced winter chill as a means to improve the reproductive potential of late day-neutral strawberry cultivars. *Eur. J. Hortic. Sci.* **2019**, *84*, 20–23. [CrossRef]
6. Melis, P. Manipulation of Everbearers to Steer Harvest Peaks and Yielding. International Soft Fruit Conference's-Hertogenbosch, Holland. 2020. Available online: http://goodberry-eu.eu/fileadmin/goodberry/appliedpublications/3737 (accessed on 10 November 2021).
7. Sønsteby, A.; Woznicki, T.L.; Heide, O.M. Effects of runner removal and partial defoliation on growth and yield performance of 'Favori' everbearing strawberry plants. *Horticulturae* **2021**, *7*, 215.
8. Sønsteby, A.; Heide, O.M. Long-day control of flowering in everbearing strawberries. *J. Hortic. Sci. Biotec.* **2007**, *82*, 875–884. [CrossRef]
9. Sønsteby, A.; Heide, O.M. Quantitative long-day flowering response in the perpetual-flowering F1 strawberry cultivar Elan. *J. Hortic. Sci. Biotec.* **2007**, *82*, 266–274. [CrossRef]
10. Samad, S.; Butare, D.; Marttila, S.; Sønsteby, A.; Khalil, S. Effects of Temperature and Photoperiod on the Flower Potential in Everbearing Strawberry as Evaluated by Meristem Dissection. *Horticulturae* **2021**, *7*, 484. [CrossRef]
11. Gallace, N.; Lieten, P. Winter manipulation of photoperiod and chill to enhance the perpetual flowering nature of 'Soprano'. *Acta Hortic.* **2021**, *1309*, 333–340. [CrossRef]
12. Rivero, R.; Sønsteby, A.; Solhaug, K.A.; Heide, O.M.; Remberg, S.F. Effects of temperature and photoperiod on photosynthesis in everbearing strawberry. *Acta Hortic.* **2021**, *1309*, 379–386. [CrossRef]

Article

Foliar Application of Some Macronutrients and Micronutrients Improves Yield and Fruit Quality of Highbush Blueberry (*Vaccinium corymbosum* L.)

Zofia Zydlik [1], Piotr Zydlik [2,*], Nesibe Ebru Kafkas [3], Betul Yesil [3] and Szymon Cieśliński [1]

[1] Department of Ornamental Plant, Dendrology and Pomology, Faculty of Agronomy, Horticulture and Bioengineering, Poznanń University of Life Sciences, 60-637 Poznan, Poland; zofia.zydlik@up.poznan.pl (Z.Z.); szym.cieslinski@gmail.com (S.C.)

[2] Department of Entomology and Environment Protection, Faculty of Agronomy, Horticulture and Bioengineering, Poznanń University of Life Sciences, 60-637 Poznan, Poland

[3] Department of Horticulture, Faculty of Agriculture, University of Çukurova, Sarıçam, Adana 01330, Turkey; ebruyasakafkas@gmail.com (N.E.K.); betulyesillll@gmail.com (B.Y.)

* Correspondence: piotr.zydlik@up.poznan.pl

Abstract: Foliar fertilization makes it possible to quickly provide plants with essential nutrients, mainly micronutrients, which can significantly improve the quality of yields. The aim of this study was to evaluate the effect of foliar fertilization with fertilizers containing calcium and microelements on yielding and fruit quality of highbush blueberry (*Vaccinium corymbosum* L.). A two-year study was carried out in western Poland in an experimental highbush blueberry production plantation. During the growing season the bushes were treated several times with the following foliar fertilizers: Armurox, BioCal, and Stymjod. The experiment assessed bush growth vigor, yield, fruit quality characteristics, sugar, organic acid, and health-promoting substance content. It was found that as a result of fertilizing highbush blueberry bushes with foliar fertilizers, the leaf blade area and plant yield increased significantly. The fruits collected from those bushes were characterized by a higher mass, firmness, and TSS content. This also applies to blueberry fruit after storage. Foliar fertilization had no significant effect on the content of chlorophyll a and b in the leaves of northern highbush blueberry, on fruit coloration, the content of sugars, ascorbic and citric acids, and the phenolic compounds in them.

Keywords: highbush blueberry; foliar fertilization; yield; fruit firmness; extract content (TSS); organic acids; sugars; phenolic compounds

Citation: Zydlik, Z.; Zydlik, P.; Kafkas, N.E.; Yesil, B.; Cieśliński, S. Foliar Application of Some Macronutrients and Micronutrients Improves Yield and Fruit Quality of Highbush Blueberry (*Vaccinium corymbosum* L.). *Horticulturae* **2022**, *8*, 664. https://doi.org/10.3390/horticulturae8070664

Academic Editor: Esmaeil Fallahi

Received: 1 July 2022
Accepted: 16 July 2022
Published: 20 July 2022

Publisher's Note: MDPI stays neutral with regard to jurisdictional claims in published maps and institutional affiliations.

1. Introduction

The main blueberry species grown on a mass scale is highbush blueberry (*Vaccinium corymbosum* L.), in Poland it also known as "American blueberry". According to statistics from the Food and Agriculture Organization of the United Nations (FAOSTAT), in 2020, its largest producers worldwide were the United States and Canada (approximately 35% and 17% of world production, respectively). There is a growing worldwide interest in the cultivation of northern highbush blueberry. While in the mid-1960s the global production of this fruit was approximately 33 thousand tons, in 2020 it was already approximately 850 thousand tons. In addition, in Polish fruit farming over the last few decades, highbush blueberry has been one of the fastest growing products [1]. According to FAOSTAT, in 2020, domestic production of this fruit was around 55 thousand tons (6.5% of global production), which places Poland 6th in the world.

The high nutritional and health-promoting properties of highbush blueberry are one of the reasons for its popularity. Its fruits contain vitamins A, B1, B2, and B3 as well as phosphorus, calcium, sodium, folic acid, and phytoestrogens. Blueberries are an excellent source of health-promoting compounds, mainly polyphenols [2–4]. One serving of blueberries, similar to a serving of cranberries or grapes, provides the body with 200–400 mg of polyphenols [5], which may be crucial in reducing the risk of developing type 2 diabetes [6,7] or other chronic diseases [8].

Shrubs of the Vaccinium genus, from which highbush blueberry originates, grow wild in soils with low nutrient content. This is why the fertilization requirements of highbush blueberry are relatively low in comparison to other fruit farming species [9]. However, high yields depend on the use of mineral fertilization to maintain the sufficiently high microelement and macroelement contents of the soil. Soil fertilization is common and effective for nutrients required in large quantities. However, doing so not only increases the number of nutrients in the soil but also changes the soil structure, enzymatic activity, and diversity of soil microorganisms [10]. Nitrogen fertilizers, in particular, can have a limiting effect on soil fungal communities by lowering soil pH. High fertilizer doses accelerate the vegetative growth of blueberry, thus reducing their yield [11].

In addition to standard soil fertilization, especially under intensive cultivation conditions, plants sometimes require additional nutrient-supplementing treatments during the growing season, i.e., foliar fertilization. It is particularly recommended when macroelements—and especially microelements—must be provided to plants. Micronutrients are part of most enzymes regulating biochemical and physiological processes in plants, and their deficiency may lead to disorders of these processes [12]. In blueberry cultivation, foliar calcium fertilization is particularly recommended [13]. This is because the blueberry is a specific species that requires a low pH soil [14], the optimum value of which should be between 4.5 and 4.8 (pH H_2O). In such a substrate, high concentrations of Al^{3+}, Fe^{2+}, and Mn^{2+} inhibit calcium ion absorption. Apart from the standard macronutrients (the aforementioned calcium), foliar fertilizers can contain ingredients that are difficult to extract from soil (e.g., silicon) or that are present in trace amounts (e.g., iodine). Both iodine and silicon activate plants' natural defense mechanisms, allowing them to mitigate the effects of stress [15]. An advantage of foliar fertilization is the rapid utilization by plants of nutrients supplied in this way, which has a positive effect on yield and fruit quality. Maintaining high quality after harvest and extending the storage life of the fruit is particularly important in the case of highbush blueberry, mainly due to the ongoing global fruit overproduction of this species. Overproduction is facilitated by, among other things, high purchase prices combined with production automation and mechanization [16]. It is assumed that properly conducted foliar fertilization can improve blueberry fruit quality both after harvest and after storage.

The aim of the study was to assess the effect of foliar fertilization with preparations containing macronutrients and micronutrients on the amount and quality of yield of northern highbush blueberry (*Vaccinium corymbosum* L.).

2. Materials and Methods

2.1. Study Sites and Experimental Design

The study was conducted between 2020 and 2021 at the experimental station of the Poznań University of Life Sciences in western Poland (52°31′ N; 16°38′ E). The research object covered highbush blueberry bushes of the Bluecrop cultivar growing on an experimental production plantation at a distance of 2.5 × 1.5 m (2667 pcs. per ha). The bushes grew in podzolic soil, formed from loamy sands with a slightly acid reaction (pH 5.6), humus content of 4.78%, and salinity of 0.59 g NaCl dm^3. The content of macroelements in the soil was (in mg dm^3): N-NO$_3$—62; P—61; K—194; Ca—952; Mg—230; Cl—41. The soil nitrogen and phosphorus contents were optimal, the potassium and calcium contents were very high, and the magnesium contents were low.

The experiment consisted of four treatments: (1) foliar spraying with water (control combination); (2) foliar spraying with Stymjod; (3) foliar spraying with BioCal; (4) foliar spraying with Armurox. They were established in a randomized block design in four plots, where one plot consisted of 6 bushes. The total number of plants used in the experiment was 96.

In each year of the experiment, the highbush blueberry bushes were fertilized with foliar fertilizers three times (i.e., during flowering, fruit setting, and ripening) in a total dose of 4 L ha for Stymjod and Armurox and 1.5 L ha for BioCal. The BioCal preparation used in the experiment was a liquid foliar fertilizer containing calcium in the form of water-soluble calcium oxide at 8% *w/w* and water-soluble zinc at 3% *w/w*. Another fertilizer, Armurox, contained silicon oxide (8%), free amino acids (3%), and total nitrogen (1%). Armurox forms a physical barrier under the plant cuticle and activates plants' endogenous defense mechanisms. Stymjod is a liquid fertilizer in the form of a concentrate, produced in nanotechnology using the cold plasma effect. It contains the optimum macronutrient composition for plants (i.e., N—6.3%; P—4.58%; K—6.42%; Mg—1.69%; S—1.6%), micronutrients (i.e., B—0.46%; Cu—0.17%; Fe—0.14%; Mn—0.16%; Mo—0.028%; Zn—0.42%) as well as humic and amino acids, increasing the resistance of plants to unfavorable environmental conditions. Information regarding the composition of the preparations was obtained from their manufacturer. During the vegetation period, standard agrotechnical practices recommended for highbush blueberry cultivation were applied. In the spring period (March–April) of each year of the study and for the highbush blueberry bushes in all treatments, the soil was fertilized with nitrogen at 50 kg ha with the addition of ammonium sulphate ($(NH_4)_2SO_4$) at 20% N and 24% S.

Climatic conditions (mean monthly and mean annual air temperature and precipitation) were evaluated on the basis of measurements taken at the meteorological station located at the place of the experiment. Spring 2020 was relatively cool (especially May) with numerous frosts (down to −8 °C). Summer saw frequent prolonged water shortages (Table 1). Compared to the multiyear mean measurements, the total precipitation during the growing season in 2020 was approximately 70 mm lower. The spring of 2021 was also relatively cold. During the summer months, there was high intensity precipitation. Over the entire growing season of that year, the amount of precipitation was approximately 25 mm higher compared to the multiyear average (Table 1).

Table 1. The course of climatic conditions in the years 2020 and 2021.

Years	Months							Sum
	IV	V	VI	VII	VIII	IX	X	
				Precipitation (mm)				
2020	42.6	49.4	48.6	78.2	57.4	41.6	53.2	279.8
2021	29.2	36.8	78.6	96.7	43.8	60.7	32.4	378.2
1982–2012 [1]	28.0	48.0	63.5	78.8	61.9	41.0	32.0	353.2
				Air temperature (°C)				
2020	8.9	11.6	18.2	18.5	20.3	15.1	10.5	103.1
2021	6.2	12.3	20.1	19.0	17.5	16.4	9.1	100.6
1982–2012 [1]	9.3	14.6	17.2	19.5	18.9	14.1	9.0	102.6

[1]—average.

2.2. Measurements and Observations

In the experiment, the following measurements and analyses were performed: blueberry leaves (i.e., leaf blade area and chloroplast pigment content), content of macroelements (i.e., N-NO$_3$; P$_2$O$_5$; K$_2$O; CaO; MgO), and microelements (i.e., Zn; Cu; Mn; B) of leaves and fruit; yield level (kg per bush); fruit weight as well as their height and width; fruit coloration; their quality parameters (i.e., firmness, extract, organic acid, and sugar contents); content of phenolic compounds and anthocyanins in fruit.

2.2.1. Measurements and Analyses of Blueberry Leaves

In the experiment, the highbush blueberry leaf area was measured. For this purpose, in both years of the study, 50 leaves were randomly collected from each treatment. They were collected from the central part of shoots in the second half of July, at the time recommended for assessing the nutritional status of blueberries [17]. The collected leaves were scanned, and then the area of their blades was calculated using DigiShape 1.9 software (ver.1.9.19, Cortex Nowa, Bydgoszcz, Poland).

The content of chloroplast pigments, chlorophyll a and b, and carotenoids (in mg kgf.m.) was determined in 2020 in 25 sample leaves from the treatment. They were collected in the summer, after the last fruit harvest. Determinations of chloroplast pigments were performed using the extraction method according to Hiscox and Israel [18]. Leaf blade fragments weighing 0.5 g were cut with a cork borer. The prepared material was covered with 5 mL of dimethyl sulfoxide and placed in a water bath at 65 °C. After 20 min, using a spectrophotometer, absorbance was measured at 470 nm for β-carotene, 645 nm for chlorophyll b, and 663 nm for chlorophyll a. The calculations were based on the equation:

- Chlorophyll a = $(12.7 \times A663 - 2.7 \times A645) \times V \times (1000 \times W)^{-1}$;
- Chlorophyll b = $(22.9 \times A645 - 4.7 \times A663) \times V \times (1000 \times W)^{-1}$;
- Sum of chlorophyll a + b = $(20.2 \times A645 + 8.02 \times A663) \times V \times (10{,}000 \times W)^{-1}$.

  where V—total extract volume cm^3; W—sample weight in g.
  Carotenoid contents in fruits (in mg kg f.m.) were calculated based on the equation:

- Carotenoids = $(1000 \times A470 - 1.9 \text{ chlorophyll a} - 63.14 \text{ chlorophyll b}) \times 214^{-1}$.

2.2.2. Macro- and Microelement Contents in Leaves and Fruits

Analyses of the macroelement and microelement contents in the leaves and fruits of highbush blueberry were performed in 2021. For this purpose, 200 leaves and 100 fruits were collected from each combination, which were then dried in the dryer (Binder, Tuttlingen, Germany) (at 45–50 °C), ground, and mineralized in the presence of sulphuric acid in a mineralizer. The content of N in plant material was determined by the distillation method according to Kjeldahl; P by the vanadomolybdenum method; K, Ca, and Mg by the atomic absorption method on Zeiss Jena AAS-5 apparatus (Oberkochen, Germany). All determinations were made in the laboratory of the Chemical and Agricultural Station, in accordance with the Polish Standard, using certified reagents.

2.2.3. Yielding and Quality of Fruits

In 2020 and 2021, fruit for analysis was harvested three times: at the end of July (first date), in early August (second date), and in the second half of August (third date). The yield was determined on the basis of the weight of ripe fruit collected from a bush (kg per bush). To assess average fruit weight, 100 fruits were randomly collected from each treatment after each harvest and weighed with an accuracy to 0.01 g. Fruit height and width (mm) were measured for a sample of 100 fruits from each treatment using an electric caliber. The extract content (TSS—total soluble solids) was measured using the PR-101a refractometer (Atago Co., Ltd., Fukaya-shi, Japan) for a sample of 100 fruits from each treatment. The values are expressed in %. Fruit firmness was assessed for the same sample (g mm^{-1}). Measurements were taken using a firmness tester "Fruit Pressure Tester mod. 327" by Facchini, Alfonsine, Italy), mounted on a stand. This test is known as the Magness–Taylor test and consists of piercing the fruit flesh with a 1.5 mm pin. In both years of the study, the post-harvest stability of the fruits was also evaluated by analyzing their firmness and extract after 6 days of storage at 4–6 °C.

For the assessment of fruit coloration, a Minolta colorimeter was used in the Lab color space, based on the so-called trichromatic theory of color vision. Coloration was expressed in the color space of L * a * b *, where L * denotes the brightness from black (0) to white (100), a * the color from green (−60) to red (60), and b * the color from blue (−60) to yellow (60). Measurements were made on 50 fruits from one treatment. Four measurements were made on one fruit and then the obtained results were averaged.

The content of organic acids (i.e., ascorbic, citric, and malic acids), sugars (i.e., total, fructose, glucose, and sucrose), phenolic compounds, and anthocyanins in the blueberry fruit was tested in 2020 on a sample of 100 fruits randomly harvested from each treatment and each of the three harvest dates.

The HPLC method developed by Bozan et al. [19] was used to determine the organic acid content. For organic acids extraction, 0.25 g of the sample and 4 mL of 3% metaphosphoric acid was mixed. The mixture was placed in the ultrasonic water bath at 80 °C for 15 min, and it was sonicated and centrifuged at 5500 rpm for 15 min. Afterward, the mixture was filtered. The extract of organic acids was analyzed using a high-performance liquid chromatographic apparatus HPLC (Shimadzu LC 20A VP, Kyoto, Japan) equipped with a UV detector (Shimadzu SPD 20A VP), and we used an 87 H column (5 μm, 300 $\times$ 7.8 mm). The identified acids were evaluated according to the relevant standard calibration curves.

Changes in glucose, fructose, sucrose, and total sugar content in homogenized blueberry samples were determined using the HPLC technique according to the method developed by Crisosto [20]. One milligram of blueberry fruit powder was added to 4 mL of ultrapure water (Millipore Corp., Bedford, MA, USA). The reaction mixture was placed in an ultrasonic bath and sonicated at 80 °C for 15 min and then centrifuged at 5500 rpm for 15 min, and it was filtered before HPLC analysis. Sugar contents were determined using HPLC (Shimadzu, Prominence LC-20A) RID (Refractive Index) detector and Coregel-87C (7.8 $\times$ 300 mm). Separations were performed at 70 °C at a flow rate of 0.6 mL min^{-1}. Elution was isocratic ultrapure water. The individual sugars were calculated according to their standards. Calibration curves of the references used were created and content was determined according to these external standard calibration curves.

The total phenolic content was evaluated using Folin–Ciocalteu reagent in the modified method of Spanos and Wrolstad [21]. Briefly, methanol extract was added on 1 g of samples. Water, Folin–Ciocalteu, and 20% sodium carbonate were added to the samples taken from the supernatant of this extract and then kept in the dark for 2 h. The absorbance of all samples were measured at 760 nm with the use of a Thermo Scientific Multiskan GO microplate spectrophotometer. Quantifications were calculated through a calibration curve daily prepared with known concentrations of gallic acid (GA) standards, and the results are expressed as milligrams of GA equivalents per 100 g of dry weight (DW) of fruits.

The pH-differential absorbance method of Wrolstad [22] was employed to quantify monomeric anthocyanin pigment content of the methanol-blueberry fruit powder extract within buffers at pH 1.0 (hydrochloric acid–potassium chloride, 0.2 mol) and 4.5 (acetate acid–sodium acetate, 1 M). A UV-spectrophotometer and 1 cm disposable cell were utilized for spectral measurements at 510 and 700 nm. Anthocyanin content was calculated as mg (cyanidin-3-glucoside).

2.3. Statistical Analysis

All results obtained were subjected to one-way classification using the STATISTICA 7 program (StatSoft, Inc., Tulsa, OK, USA). The significance of the differences between the means for individual treatments was assessed using Duncan's test for the significance level of $\alpha = 0.05$. The chemical analyses were performed in four repetitions.

3. Results and Discussion

3.1. Parameters of Highbush Blueberry Leaves

Leaf shape and size are important factors affecting plant yield. To absorb enough light energy, leaves must be as wide as possible. The blueberry leaf area in the experiment, depending on the year and treatment, varied from 7.69 cm^2 in the control treatment to 13.12 cm^2 in the treatment where Stymjod was applied (Figure 1). The leaf blade area of highbush blueberry increased significantly after foliar fertilization. In the first year of the study, BioCal and Stymjod fertilizers were the most effective in this respect, with leaf area increasing by 60% and 64%, respectively, compared to the control treatment. In 2021, the increase in the leaf blade area in the treatments with foliar fertilization compared to the control treatment was smaller, on the order of a few percentage units.

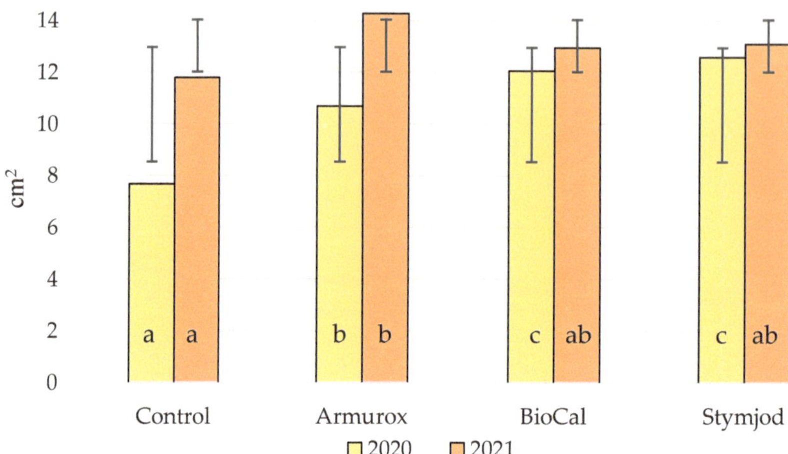

Figure 1. Effect of foliar fertilization on leaf blade area (cm^2) of highbush blueberry in 2020 and 2021. The mean values marked with the same letters do not differ significantly at a = 0.05.

Photosynthesis is the main factor responsible for the process of biomass accumulation by plants. In the experiment, there was no significant variation in the chlorophyll content of highbush blueberry leaves. Two of the three tested foliar fertilizers did not affect its content either. Only in the treatment with the bushes sprayed with the Stymjod fertilizer did the total content of chlorophyll a and b in the leaves increase significantly (Table 2). The authors also did not obtain confirmation of previous results by other researchers [23,24] showing an increase in chlorophyll content in plants growing under stress condition under the influence of silicon. In the highbush blueberry leaves sprayed with Armurox (8% Si), the chlorophyll content did not differ significantly from that in leaves of bushes from the other treatments. Carotenoids perform a protective function for chlorophyll. In the experiment, the differences in their content in the leaves of highbush blueberry foliar fertilized were recorded. The content of carotenoids in leaves from bushes treated with BioCal and Stymjod fertilizers was higher than the control.

Table 2. Content of chloroplast pigments in highbush blueberry leaves (mg kg^{-1} f.m.) in 2020.

Treatments	Chlorophyll a	Chlorophyll b	Chlorophyll a + b	Carotenoids
Control	54.56 ± 0.10a [1]	101.46 ± 0.54a	156.13 ± 0.62ab	750.30 ± 3.16a
Armurox	54.60 ± 0.08a	100.50 ± 1.03a	155.20 ± 1.09a	757.43 ± 6.61ab
BioCal	54.85 ± 0.32ab	101.65 ± 0.76ab	156.61 ± 1.08bc	790.34 ± 4.70b
Stymjod	54.95 ± 0.12b	102.76 ± 0.51b	157.82 ± 0.14c	997.42 ± 8.97c

[1] The mean values marked with the same letters did not differ significantly at a = 0.05.

3.2. Macroelements and Microelements Content of Leaves and Fruit

The mineral content of leaves is an indicator of the nutritional status of the plant. In the experiment, in the leaves of highbush blueberry, depending on the treatment, the contents of macroelements were (in % d.m.) N_{tot}—from 1.53 to 1.70; P_2O_5—0.26; K_2O—0.53; CaO—from 0.92 to 1.12; MgO—from 0.41 to 0.49 (Table 3).

Table 3. The content of macro- and microelements of highbush blueberry leaves in 2021.

Treatments	Macroelements (% d.m.)					Microelements (mg kg^{-1} d.m.)			
	Ntot	P$_2$O$_5$	K$_2$O	CaO	MgO	Zn	Cu	Mn	B
Control	1.56 ± 0.011b	0.26 ± 0.010a	0.53 ± 0.012a	1.04 ± 0.07c	0.44 ± 0.017a	7.75 ± 0.10a	2.96 ± 0.36a	213 ± 12.4d	33.60 ± 2.6c
Armurox	1.53 ± 0.019a	0.26 ± 0.017a	0.53 ± 0.009a	0.92 ± 0.04a	0.41 ± 0.021a	8.86 ± 0.35a	3.68 ± 0.54c	173 ± 18.7a	31.05 ± 1.9b
BioCal	1.67 ± 0.013c	0.25 ± 0.021a	0.52 ± 0.017a	1.12 ± 0.11d	0.46 ± 0.011a	32.0 ± 0.98c	3.24 ± 0.31b	204 ± 25.4c	30.05 ± 3.8a
Stymjod	1.70 ± 0.021d	0.26 ± 0.019a	0.53 ± 0.011a	0.96 ± 0.09b	0.49 ± 0.018a	17.0 ± 0.67b	7.67 ± 0.72d	186 ± 17.6b	49.30 ± 2.9d

The mean values marked with the same letters did not differ significantly at a = 0.05.

The literature contains numerous and often varied results of macronutrient content in highbush blueberry leaves. The contents obtained in this experiment were higher than in earlier studies by the authors of the article, where the macronutrient contents in blueberry leaves under various conditions, including replant disease [25], were analyzed, or similar to the results obtained in experiments by other researchers. These contents were (in % d.m.) N—from 1.80 to a maximum of 2.29 [17,26–28]; P—from 0.49 [17] to 0.90 [28]; K—minimum of 0.13; Mg—from 0.17 to 0.19 [17]. In the experiment, the content of N-NO$_3$ and Mg (BioCal and Stymjod fertilizers) significantly increased in the leaves of highbush blueberry after foliar fertilization. In treatment with BioCal (8% Ca), an especially high content of calcium was also found in blueberry leaves compared to the control treatment.

The contents of microelements in the leaves of highbush blueberry were (in mg kg^{-1} d.m.) Zn—from 7.75 to 32.0; Cu—0.29; K—0.53; Ca—from 0.92 to 1.12; Mg—from 0.41 to 0.49 (Table 3). In blueberry leaves from foliar-fertilized bushes, the content of Zn and Cu was significantly higher and Mn was lower than in the control treatment. The greatest differences were demonstrated for BioCal and Stymjod. It should be added that the composition of the latter one, besides amino acids, included a number of microelements. Considerable differences in Zn content were also found. The content of this element in leaves in the treatments with bushes sprayed with BioCal and Stymjod, as compared to the control, was over four times higher (7.75 and 32.0 mg kg^{-1} d.m., respectively) (Table 3). As Raj and Raj [12] pointed out, Zn is essential for plant nutrition. It participates in all major plant functions, increases nitrogen uptake and, at later stages, activates CO$_2$ fixation.

In the fruits of highbush blueberry, depending on the treatment, the contents of macroelements (in % d.m.) were N—from 0.60 to 0.76; P—0.26; K—from 0.69 to 0.80; Ca—0.15; Mg—0.08 (Table 4). The contents of macroelements in the blueberry fruit was less varied than in the leaves. The amount of Ca and Mg in fruits was identical, regardless of the treatments. In the blueberry fruit from the bushes sprayed with Armurox and Stymjod, the contents of N, P, and K were lower than in the fruit from the bushes that were not foliar fertilized (Table 4). The position of Ochmian et al. [29,30] regarding an increase in Ca and P contents in blueberry fruit as a result of spraying bushes with Ca- and P-containing fertilizers was not confirmed.

Table 4. The content of macro- and microelements of highbush blueberry fruits in 2021.

Treatments	Macroelements (% d.m.)					Microelements (mg kg^{-1} d.m.)			
	N-NO$_3$	P$_2$O$_5$	K$_2$O	CaO	MgO	Zn	Cu	Mn	B
Control	0.74 ± 0.020c	0.26 ± 0.030b	0.80 ± 0.12c	0.15 ± 0.07a	0.08 ± 0.021a	6.13 ± 0.45b	2.94 ± 0.57b	30.4 ± 2.45d	5.28 ± 0.74c
Armurox	0.64 ± 0.017b	0.23 ± 0.027a	0.69 ± 0.09a	0.15 ± 0.03a	0.08 ± 0.016a	5.40 ± 0.26a	2.48 ± 0.23a	26.8 ± 1.98b	4.18 ± 0.95a
BioCal	0.76 ± 0.061d	0.26 ± 0.019b	0.74 ± 0.15b	0.15 ± 0.02a	0.08 ± 0.009a	24.7 ± 1.34d	2.97 ± 0.19c	21.3 ± 0.86a	4.43 ± 0.56b
Stymjod	0.60 ± 0.019a	0.26 ± 0.036b	0.74 ± 0.17b	0.15 ± 0.03a	0.08 ± 0.010a	6.52 ± 0.29c	4.04 ± 0.36d	28.9 ± 1.73c	8.80 ± 1.03d

The mean values marked with the same letters did not differ significantly at a = 0.05.

In the fruits of highbush blueberry, depending on the treatment, the contents of microelements were (in mg kg^{-1} d.m.) Zn—from 5.40 to 24.7; Cu—from 2.48 to 4.04; Mn—from 21.3 to 30.4; B—from 4.18 to 8.80 (Table 4). Significant differences in micronutrient contents occurred in the treatment with Stymjod. In fruits from the bushes sprayed with this preparation, significantly higher contents of Zn, Cu, Fe, and B were found in comparison to the control. It should be added that the composition of this fertilizer included a whole range of microelements. The content of Zn in blueberry fruit increased evenly several times. This was true for the treatment with the foliar application of BioCal.

3.3. Yielding of Blueberry Bushes

A high yield is the determining factor for profitable production. In the first year of the experiment, spraying highbush blueberry bushes with each foliar fertilizer resulted in a significant increase in their yield. The difference compared to the control treatment ranged from approximately 30% (Stymjod) to over 40% (Armurox) (Figure 2). In the second year of the experiment, the yielding of blueberry bushes in each treatment was higher than in the previous year. However, in that period, no significant effect of foliar fertilization on the yielding of highbush blueberry bushes was found.

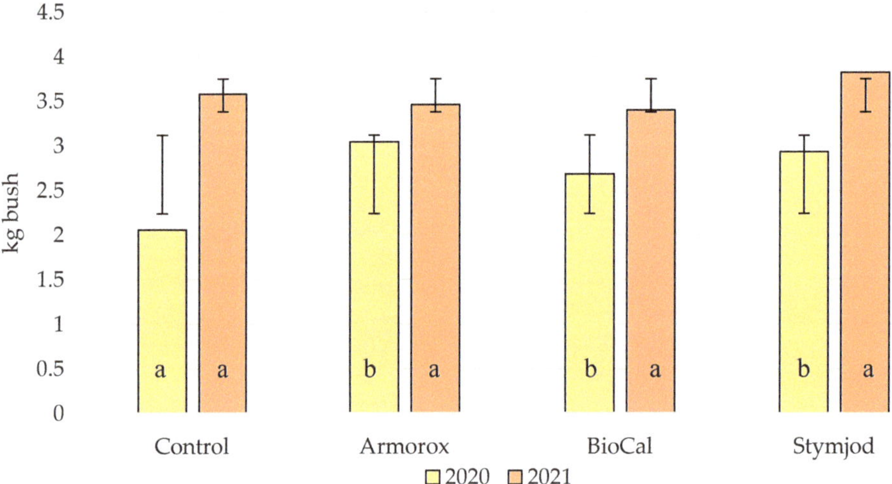

Figure 2. Yielding of blueberry bushes (kg bush) in 2020 and 2021. The mean values marked with the same letters did not differ significantly at a = 0.05.

The high effectiveness of foliar fertilization in increasing the yield of highbush blueberry bushes only in 2020 may be due to the fact that in that period there were hardly favorable weather conditions for cultivation. For example, the amount of precipitation was low (Table 1), which resulted in long periods of drought. Under such conditions, the effectiveness of foliar fertilization may have increased. The characteristics of the foliar fertilizers used in the experiment show that one of their important properties is the mitigation of stress conditions.

3.4. Biometric and Quality Parameters of Highbush Blueberry Fruit

Biometric parameters (i.e., height and width) and fruit weight are important physical indicators affecting fruit attractiveness. Fruit size is vital for economic reasons. Large fruit is much easier to sell than smaller fruit. In the experiment, the smallest highbush blueberry fruit was collected from the bushes that were not foliar fertilized (control). In the experiment, the range of fruit weight measurements was quite considerable, from 1.39 g in the control treatment to 2.19 g in the treatment with BioCal fertilizer application in 2021 (Table 5). The available literature shows that such differences in highbush blueberry fruit weight were demonstrated in the works of many authors. Fluctuations may range from 1.12 to 2.11 g according to Aliman et al. [31]; from 1.4 to 2.4 g according to Correia et al. [32];

from 1.76 to 1.94 g according to Zorenc et al. [33]; from 1.29 to 1.80 g. These should be explained, among other things, by varietal differences or weather conditions.

Table 5. Effect of foliar fertilization on highbush blueberry fruit weight and biometric parameters in 2020 and 2021.

Treatments	Weight (g)		Height (mm)		Width (mm)	
	2020	2021	2020	2021	2020	2021
Control	1.39 ± 0.46a [1]	1.63 ± 0.43a	9.91 ± 1.12a	9.98 ± 0.97a	13.30 ± 1.44a	14.29 ± 1.34a
Armurox	1.92 ± 0.35b	2.14 ± 0.37b	10.85 ± 1.13c	10.81 ± 0.80b	15.46 ± 0.97d	15.73 ± 1.09b
BioCal	1.89 ± 0.45b	2.19 ± 0.43b	10.80 ± 0.85c	11.17 ± 0.75c	15.19 ± 1.38c	15.92 ± 1.28b
Stymjod	1.91 ± 0.39b	2.14 ± 0.29b	10.58 ± 0.86b	11.32 ± 0.62c	14.94 ± 1.23b	15.75 ± 0.94b

[1] The mean values marked with the same letters did not differ significantly at a = 0.05.

In both years of the study, the bushes sprayed with foliar fertilizers produced fruit with significantly higher weight and biometric parameters (i.e., height and width) compared to the bushes without foliar fertilization. All three foliar fertilizers were equally effective in this respect. The effectiveness of foliar fertilization with calcium-containing fertilizers in increasing highbush blueberry fruit weight was also reported by Ochmian et al. [29].

With the growing health consciousness of the public, consumer attention is increasingly focused on the nutritional quality of foodstuffs. Fruit firmness is considered one of the most important quality characteristics, determining, among other things, its commercial value. The resistance of fruit to mechanical damage largely depends on the condition of cell membranes. Silicon and calcium play a leading role in their stabilization. The latter determines the suitability of fruit for transport and storage by stabilizing pectins that form the central lamina of the cell walls. As Ochmian and Kozos [29] pointed out, an increase in the calcium content of fruit increases its firmness. The blueberry fruit firmness shown in the experiment ranged from 176.0 g mm^{-1} in 2020 to 293.55 g mm^{-1} a year later (Table 6). In both years of the study, fruits with the lowest firmness were harvested from bushes in the control treatment, i.e., without foliar fertilization. As a result of foliar spraying with each of the three fertilizers, the firmness of blueberry fruit increased significantly. Compared to the control treatment, the difference ranged from approximately 32% in 2020 (Armurox) to almost 50% in 2021 (Stymjod) (Table 6). The effectiveness of BioCal in this respect, despite its significant calcium content, was not the highest. This may have been due to the difficult transport of this element into the cell during foliar spraying [34].

Table 6. Effect of foliar fertilization on blueberry fruit firmness (g mm^{-1}) and TSS content (%) in 2020 and 2021 (TSS—total soluble solids).

Treatments	2020		2021			
			Firmness		TSS	
	Firmness	TSS	H [1]	S [2]	H [1]	S [2]
Control	176.00 ± 30.59a [3]	14.29 ± 2.75a	196.95 ± 21.29a	176.6 ± 25.85a	11.85 ± 1.51a	12.75 ± 1.29a
Armurox	232.53 ± 41.62b	13.20 ± 1.62a	264.20 ± 34.81b	242.3 ± 26.35b	14.43 ± 1.44c	14.46 ± 1.43b
BioCal	242.40 ± 34.66c	13.88 ± 1.35a	285.20 ± 41.92c	241.4 ± 20.13b	13.96 ± 1.73c	13.90 ± 1.47a
Stymjod	245.07 ± 36.42c	14.19 ± 1.71a	293.55 ± 38.02d	249.10 ± 22.21c	14.54 ± 1.94c	16.46 ± 1.54b

[1] H—fruit after harvest. [2] S—fruit after storage. [3] The mean values marked with the same letters did not differ significantly at a = 0.05.

The dessert value of the fruit is determined by its high extract content (TSS) [35]. Similar to firmness, the TSS value in highbush blueberry fruit was the lowest in the control treatment (11.85%) and the highest (14.54%) in the treatment where Stymjod was applied.

This TSS content in fruit was higher than in the experiment of Shevchuk et al. [36] and similar to the results obtained by Zenkov and Pinchykov [37], from 13 to 15.3%, or by Celik et al. [38]—approximately 13.3%. In 2020, there was no variation in extract content in blueberry fruit. The causal factor in this case could have been weather conditions. As Ersoy et al. [39] pointed out, the physicochemical properties of fruit can be affected by factors such as variety, cultivation method, and weather conditions. The TSS content in blueberry fruit varied in 2021. The bushes fertilized with the foliar fertilizer had a higher extract content than the bushes without foliar fertilization. Differences ranged from a few percent (BioCal) to over 20% (Stymjod and Armurox) (Table 6).

Producers not only want to maintain the high quality of their fruit after harvest, but also after storage. The experiment showed that regardless of the treatment, the firmness of highbush blueberry fruit after six days of storage was lower than after harvest. The difference ranged from a few percent for the treatment with the application of Armurox to 17–18% for treatments with BioCal and Stymjod (Table 6). The firmness of stored blueberry fruit from the foliar-fertilized bushes was significantly higher than that of fruit from the bushes to which foliar fertilizers were not applied. The Stymjod fertilizer was the most effective in this respect. The firmness of stored blueberry fruit from this treatment (249.10 g mm^{-1}) was over 40% higher than that of the control (176.6 g mm^{-1}). Differences in the firmness of stored blueberry fruit from bushes to which Armurox and BioCal were applied were approximately 37% (Table 6).

The extract content of highbush blueberry fruit after storage, regardless of the treatment, was higher than after harvest. The effect of foliar fertilizers on the extract content of blueberry fruit after storage was less than on their firmness. The differences ranged from less than one percent in the treatment with Armurox to 13% in the treatment with Stymjod. After application of the latter one, the TSS content in stored highbush blueberry fruit (16.46%) was about 30% higher than in the control treatment (12.75%) (Table 6).

The color of the fruit is an important indicator of its maturity and harvest date [40]. The foliar fertilizers used in the experiment modified blueberry fruit coloration to a small extent. L and b values in post-harvest fruit were not significantly different (Table 7). Blueberry fruits sprayed with Stymjod were the brightest (L = 26.65), while those from the treatment without foliar fertilization were the darkest (L = 28.08). However, statistical analysis did not show the significance of differences between the measurements. Greater changes in fruit coloration occurred after storage. The fruits with the darkest color were those of the control treatment (L 27.76), and those with the lightest color were those of the control treatment (L value 26.38) (Table 7).

Table 7. Highbush blueberry fruit coloration in 2021.

Treatments	After Harvest			After Storage		
	L	a	b	L	a	b
Control	28.08 ± 3.13a [1]	0.86 ± 1.58b	5.15 ± 0.44a	27.76 ± 3.27c	1.29 ± 0.25b	4.79 ± 0.60a
Armurox	27.38 ± 2.73a	0.27 ± 1.36a	4.93 ± 0.61a	27.43 ± 3.56bc	0.87 ± 0.17a	4.16 ± 0.21a
BioCal	28.08 ± 3.22a	1.03 ± 1.66b	5.38 ± 0.53a	26.38 ± 3.11a	1.14 ± 0.13ab	4.68 ± 0.25a
Stymjod	26.65 ± 2.69a	0.12 ± 1.42a	5.42 ± 0.61a	26.65 ± 3.15ab	0.99 ± 0.12ab	4.54 ± 0.34a

[1] The mean values marked with the same letters did not differ significantly at a = 0.05.

3.5. Sugar Content and Acidity of Highbush Blueberry Fruits

Consumers initially make their choice of fruit based on its appearance and are then guided by other sensory factors—taste and sugar content [41]. It is the amount and ratio of soluble sugars and organic acids in the fruit that determines its flavor [42]. Sugars are the basic products of photosynthesis, providing energy for all biochemical processes in the cells [43]. In highbush blueberry fruits, it is the sugars that determine their organoleptic quality [44], in particular the taste [45]. The total sugar content of fruit can depend on a number of factors: plant species and cultivar, age of the plant, soil properties, climatic conditions, agrotechnical practices, and occurrence of biotic and abiotic stress [46,47]. In the experiment, the total sugar content varied from 12.96 to 15.22 mg 100 g^{-1} depending on the treatments (Table 8). This was more than in the experiment of Aliman et al. [31], from 9.73 to 9.94 mg 100 g^{-1}, and comparable to the results obtained by Kirin et al. [48], from 10.15 to 14.8 mg 100 g^{-1}.

Table 8. Effect of foliar fertilization on the sugar content of highbush blueberry fruits in 2020.

Treatments	Fructose	Glucose	Sucrose	Total Sugar
Control	6.03 ± 1.21b [1]	9.48 ± 1.17b	0.19 ± 0.02a	15.22 ± 1.38b
Armurox	4.89 ± 0.11a	7.92 ± 0.70a	0.18 ± 0.02a	12.96 ± 1.03a
BioCal	5.50 ± 0.61ab	8.83 ± 0.13ab	0.18 ± 0.07a	14.51 ± 0.76b
Stymjod	5.39 ± 0.79ab	8.95 ± 0.86ab	0.19 ± 0.42a	14.53 ± 0.98b

[1] The mean values marked with the same letters did not differ significantly at a = 0.05.

The experiment did not show any significant differentiation in total sugar content in highbush blueberry fruits harvested from bushes fertilized with and without foliar fertilizers. Blueberry bushes sprayed with BioCal and Stymjod yielded fruit with a sugar content comparable to that of the control treatment. Regardless of the treatment, glucose was the most abundant sugar and sucrose the least (Table 8). Glucose is considered the most important sugar found in blueberry fruits [49]. The content of fructose and glucose in fruit harvested from bushes sprayed with these fertilizers was also not significantly different from their content in the control treatment.

In the experiment, the authors detected little variation in the organic acid content of highbush blueberry fruit in different treatments. Foliar fertilization had very little effect on this parameter. The content of ascorbic and citric acids in all treatments was not significantly different. The greatest differences were in the content of malic acid. In the treatment with the foliar application of Stymjod, its amount (1.00 mg 100 g^{-1}) was approximately 30% lower than in the treatment without foliar application (1.42 mg 100 g^{-1}) (Table 9). The differences in malic acid content in fruit from bushes treated with BioCal and Armurox were not significant.

Table 9. Effect of foliar fertilization on organic acids content (mg 100 g^{-1}) of highbush blueberry fruits in 2020.

Treatments	Ascorbic Acid	Citric Acid	Malic Acid
Control	44.32 ± 6.16ab [1]	9.12 ± 1.73a	1.42 ± 0.39b
Armurox	60.12 ± 3.36b	8.75 ± 2.42a	1.08 ± 0.16ab
BioCal	36.58 ± 2.99a	8.38 ± 2.46a	1.06 ± 0.31ab
Stymjod	51.99 ± 3.34b	9.65 ± 2.37a	1.00 ± 0.52a

[1] The mean values marked with the same letters did not differ significantly at a = 0.05.

Ascorbic acid refers to bioactive compounds that act as antioxidants in the human body [38]. Its varying content in blueberry fruit is a varietal characteristic [50]. Fluctuations in ascorbic acid content depending on habitat conditions can range from 6 to 162 mg 100 g^{-1} [32]. The content of ascorbic acid in blueberry fruits in the experiment varied from 35.58 mg 100 g^{-1} (treatment with BioCal fertilizer) to 60.12 mg 100 g^{-1} (treatment with Armurox fertilizer) (Table 9). This was more than in the experiment by Rupasova et al. [51] at 11.8 mg 100 g^{-1} and comparable to the results of Zenkova and Pinchykova's [37] experiment at 60.5–72.2 mg 100 g^{-1}. The content of the other two acids—citric and ascorbic acids—in highbush blueberry fruit did not differ significantly regardless of the treatment (Table 9).

3.6. The Content of Health-Promoting Compounds in Highbush Blueberry Fruit

Phenolic compounds constitute the largest group of antioxidants that inhibit the oxidation reaction and reduce the amount of so-called oxygen free radicals, which reduces the risk of, for example, coronary heart disease [52]. They can be found in numerous plants. Their content in the fruit depends, among other things, on the variety, coloration and degree of ripeness. In the experiment, the total content of phenolic compounds ranged from 1606.18 mg GA 100 g^{-1} d.m. (BioCal treatment) to 1779.63 mg GA 100 g^{-1} d.m. (Armurox treatment) (Table 10). These values are similar to the results obtained by Sadowska et al. [53] at 1768.63 mg GA 100 g^{-1} d.m phenolic compounds in blueberry fruit and 1776.18 mg GA 100 g^{-1} d.m. in chokeberry fruit [54].

Table 10. Effect of foliar fertilization on the content of health-promoting compounds in blueberry fruit in 2020.

Treatments	Total Phenolic Compounds (mg GA 100 g^{-1} d.m.)	Anthocyanins (mg 100 g^{-1} d.m.)
Control	1716.25 ± 103.69a [1]	38.76 ± 3.94b
Armurox	1779.63 ± 139.09a	32.46 ± 1.67a
BioCal	1606.18 ± 220.86a	32.32 ± 1.45a
Stymjod	1631.84 ± 190.88a	32.93 ± 1.94a

[1] The mean values marked with the same letters did not differ significantly at a = 0.05.

The authors of the experiment did not detect any significant differentiation in the total content of phenolic compounds in the fruit of highbush blueberry from bushes fertilized with and without foliar nutrition. According to Ochmian et al. [30], increasing the doses of foliar fertilizers can even lead to a reduction in their content in blueberry fruit. The anthocyanin content of blueberry fruit depends on variety, size of the fruit, degree of ripeness, and climatic conditions and can vary from 25 to 490 mg 100 g^{-1} d.m. [55]. In the experiment, the anthocyanin content of highbush blueberry fruit ranged from 32.32 to 38.76 mg 100^{-1} d.m. (Table 10). These results were higher than in Shevchuck et al.'s [36] experiment, which were from 35 to 68 mg 100^{-1} d.m. in blueberry fruit, and much lower than in cranberry fruit at 370 mg 100^{-1} d.m. [52]. In the experiment, the authors did not show any significant effect of foliar fertilization on anthocyanin content in highbush blueberry fruit. In fruit from bushes in treatments with such fertilization, the anthocyanin content was lower than from bushes in the control (Table 10).

3.7. Impact of Harvesting Dates on the Tested Parameters of Highbush Blueberry

Fruit harvest dates, season, temperature differences, and amount of precipitation have an impact on quality features of blueberry fruit, including its firmness, weight, sugar and acid content [56,57]. The experiment showed such significant effect of fruit harvesting dates on a number of the tested parameters. This concerns, among other things, the yield of highbush blueberries. The highest yield, both in the first and second years of the experiment, was harvested on the second date, in early August (1.08 and 1.67 kg per bush, respectively), and the lowest on the third date, in mid-August (0.45 and 1.02 kg per bush) (Figure 3).

The biometric parameters of highbush blueberry fruit also differed significantly depending on the harvest date. Regardless of the study year, fruits with significantly higher size parameters were collected from blueberry bushes at the beginning of the harvest than at the other dates. This applied to the average weight as well as the height and width of the fruit (Table 11). In the last third of the harvesting season, especially in 2020, the fruit with the lowest weight and the smallest height and width were collected from the blueberry bushes. In 2020, the amount of precipitation was much lower than a year later, and as Zorenc et al. [33] showed, blueberry fruit weight increases with an increase in rainfall.

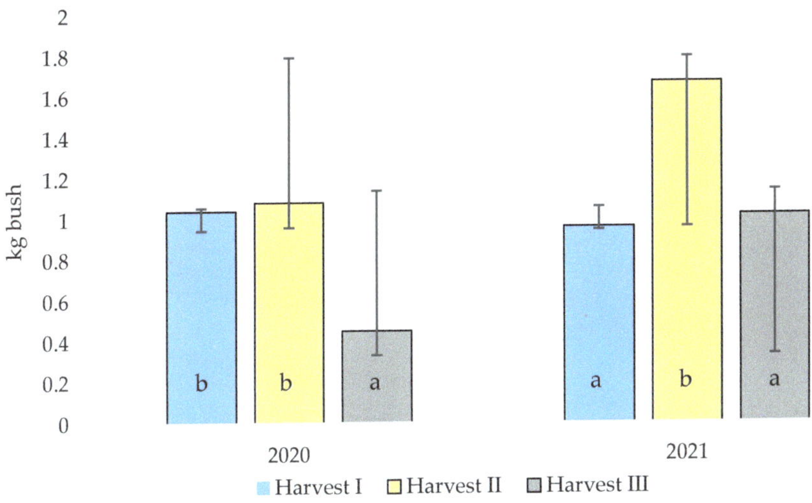

Figure 3. Impact of harvest time on the yielding of highbush blueberry bushes (kg bush) in 2020 and 2021 (harvest I—end of July; harvest II—beginning of August; harvest III—mid-August). The mean values marked with the same letters did not differ significantly at a = 0.05.

Table 11. Impact of harvest time on highbush blueberry fruit weight and their biometric parameters in 2020 and 2021.

Harvest Dates	Weight (g)		Height (mm)		Width (mm)	
	2020	2021	2020	2021	2020	2021
End of July (I)	2.11 ± 0.37c [1]	2.09 ± 0.31c	11.27 ± 0.62b	11.49 ± 0.65c	15.57 ± 1.03c	15.73 ± 1.05c
Beginning of August (II)	1.69 ± 0.30b	1.74 ± 0.29a	10.20 ± 0.47a	10.47 ± 0.67b	14.50 ± 1.02b	16.29 ± 0.98a
Mid-August (III)	1,54 ± 0.29a	1.88 ± 0.30b	1013 ± 0.82a	10.09 ± 0.68a	14.16 ± 1.08a	15.51 ± 15.51a

[1] The mean values marked with the same letters did not differ significantly at a = 0.05.

Another characteristic showing visible variation depending on the harvest dates was the quality of the fruit. In this case, there was not as much repeatability as in the analysis of yielding and fruit size. This may have been due to the mentioned variation in weather conditions occurring in the two years of the experiment (Table 1). In 2020, at the beginning of August (second harvest date), blueberry fruits were the least firm (211.8 g mm^{-1}) (Table 12). In the following research year, the firmness of fruit harvested at this date was highest (281.45 g mm^{-1}) and lowest at the end of the harvest. Greater reproducibility in both years of the study was shown with TSS content analysis in highbush blueberry fruit. In both years of the study, the TSS value was the highest (14.73 and 13.77%) in blueberry fruits harvested in early August (second term).

Table 12. Impact of harvest time on several quality parameters of highbush blueberry fruit in 2020 and 2021.

Harvest Dates	Firmness (g mm^{-1})		TSS (%)	
	2020	2021	2020	2021
End of July (I)	223.26 ± 37.67b [1]	286.30 ± 25.50b	12.81 ± 1.07a	12.65 ± 0.96a
Beginning of August (II)	211.80 ± 31.51a	281.45 ± 40.61c	14.73 ± 1.44b	13.77 ± 1.36c
Mid-August (III)	227.55 ± 25.02b	225.50 ± 27.05a	14.15 ± 1.37ab	13.17 ± 1.44b

[1] The mean values marked with the same letters did not differ significantly at a = 0.05.

In fruits harvested on the first date (end of July), the total sugar content (15.08 mg 100^{-1}) was the highest, and on the last date, mid-August, it was the lowest (13.45 mg 100^{-1}) (Table 13). The same can be said for the fructose and glucose content of fruit. The highest content of these sugars was found in fruit harvested on the first date (glucose) or the first and second dates (fructose). The content of organic acids in highbush blueberry fruit varied considerably depending on the harvest date. The contents of ascorbic (68.12 mg 100 g^{-1}), citric (12.47 mg 100 g^{-1}), and malic (1.47 mg 100 g^{-1}) acids in fruits from the third harvest were significantly higher than those from the first harvest (Table 13). These results are different from those obtained by Zorenc et al. [33]. In their experiment, at the end of the vegetation period the sugar content in highbush blueberry fruits was the highest, and the acid content-the lowest. The highest total content of phenolic compounds and anthocyanins (1762.71 mg GA 100 g^{-1} and 37.32 mg 100 g^{-1}, respectively), was detected in fruits harvested on the last date–mid-August, and the lowest (1535.89 mg GA 100 g^{-1} and 30.98 mg 100 g^{-1}) on the second date (Table 13).

Table 13. Impact of harvest time on the content of sugar, organic acids, and health-promoting compounds in blueberry fruit (harvest I—end of July; harvest II—beginning of August; harvest III—mid-August).

Studied Characteristic	Harvest I	Harvest II	Harvest III
Sugar content (mg 100 g^{-1})			
Fructose	5.97 ± 1.17b [1]	5.64 ± 0.32b	4.74 ± 0.58a
Glucose	9.29 ± 1.16b	8.57 ± 0.33a	8.53 ± 0.74a
Sucrose	0.19 ± 0.03ab	0.17 ± 0.03a	0.20 ± 0.06b
Total sugar	15.08 ± 1.41c	14.38 ± 1.01b	13.45 ± 1.33a
Organic acids (mg 100 g^{-1})			
Ascorbic acid	28.86 ± 7.51a	40.29 ± 26.92b	68.12 ± 27.57c
Citric acid	8.26 ± 1.33b	6.12 ± 1.69a	12.47 ± 2.06c
Malic acid	1.01 ± 0.42a	0.94 ± 0.29a	1.47 ± 0.29b
Health-promoting compounds			
Phenolic compounds	1751.81 ± 92.77b	1535.89 ± 199.21a	1762.71 ± 134.34c
Anthocyanins	34.06 ± 1.28b	30.98 ± 3.99a	37.32 ± 3.22c

[1] The mean values marked with the same letters did not differ significantly at a = 0.05.

4. Conclusions

In the experiment, the authors obtained confirmation of the previous assumption of a positive effect of foliar fertilization of highbush blueberry bushes on its vegetative growth, yield, and fruit quality. In treatments with the application of foliar fertilizers, the leaf area of blueberries increased significantly compared to unfertilized bushes. Fruit yields also increased by several tens of percent. The effect of foliar fertilizers on yielding, especially BioCal and Armurox, was most evident in the first year of their application, under unfavorable climatic conditions for blueberry cultivation (water deficiencies). The blueberry bushes that were foliar fertilized yielded fruit with better quality parameters than those not fertilized. This applies to average fruit weight, firmness and extract content (TSS). Foliar fertilizers also improved blueberry fruit quality after storage. Stored fruit from bushes treated with foliar fertilizers had a higher extract content and kept their firmness longer, compared to bushes without foliar fertilization.

The authors showed no significant effect of foliar fertilization on the contents of chlorophyll a and b in the leaves of highbush blueberry or on its fruit coloration, total sugar content, and ascorbic and citric acids. There was also no significant effect of the tested foliar fertilizers on the content of health-promoting compounds in highbush blueberry fruit. Yielding of highbush blueberry bushes, biometric parameters of fruit, their quality characteristics, and content of health-promoting substances significantly differed depending on the harvest date. The highest yields of blueberry fruit were harvested in early August (second term). The fruits with the highest weight and biometric parameters (i.e., height and width) were harvested at the beginning of the harvest, at the end of July. Fruits harvested on this date also had the highest total sugar content. On the other hand, in blueberry fruits from the last harvest date (mid-August), the content of all tested organic acids was the highest. The obtained results of research about foliar fertilization of blueberries can be used to increase its yield and improve fruit quality also after their storage.

Author Contributions: Conceptualization, Z.Z. and P.Z.; methodology, Z.Z. and N.E.K.; validation, P.Z. and Z.Z.; formal analysis, S.C.; investigation, Z.Z., P.Z., N.E.K. and B.Y.; data curation, P.Z.; writing—original draft preparation, Z.Z. and S.C.; writing—review and editing, P.Z.; visualization, P.Z. and N.E.K.; supervision, Z.Z.; project administration, Z.Z. and P.Z.; funding acquisition, Z.Z., B.Y. and N.E.K. All authors have read and agreed to the published version of the manuscript.

Funding: Publication was co-financed within the framework of the Polish Ministry of Science and Higher Education's program: "Regional Initiative Excellence" in the years 2019–2022 (No. 005/RID/2018/19), financing amount 12,000,000 PLN.

Informed Consent Statement: Not applicable.

Conflicts of Interest: The authors declare no conflict of interest.

References

1. Zmarlicki, K.; Brzozowski, P. *Uwarunkowania w Produkcji Agrestu, Czarnej Porzeczki i Borówki Wysokiej*; Instytut Ogrodnictwa: Skierniewice, Poland, 2016; 11p.
2. Routray, W.; Orsat, V. Blueberries and their anthocyanins: Factors affecting biosynthesis and properties. *Compr. Rev. Food Sci. Food Saf.* **2011**, *10*, 303–320. [CrossRef]
3. Norberto, S.; Silva, S.; Meireles, M.; Faria, A.; Pintado, M.; Calhau, C. Blueberry anthocyanins in health promotion: A metabolic overview. *J. Funct. Foods* **2013**, *5*, 1518–1528. [CrossRef]
4. Diaconeasa, Z.; Leopold, L.; Rugină, D.; Ayvaz, H.; Socaciu, C. Antiproliferative and antioxidant properties of anthocyanin rich extracts from blueberry and blackcurrant juice. *Int. J. Mol. Sci.* **2015**, *16*, 2352–2365. [CrossRef] [PubMed]
5. Scalbert, A.; Williamson, G. Dietary intake and bioavailability of polyphenols. *J. Nutr.* **2000**, *130*, 2073S–2085S. [CrossRef]
6. Guo, X.F.; Ruan, Y.; Li, Z.H.; Li, D. Flavonoid subclasses and type 2 diabetes mellitus risk: A meta-analysis of prospective cohort studies. *Crit. Rev. Food Sci. Nutr.* **2019**, *59*, 2850–2862. [CrossRef]
7. Johnson, S.A.; Arjmandi, B.H. Evidence for anti-cancer properties of blueberries: A mini-review. *Anti-Cancer Agents Med. Chem.* **2013**, *13*, 1142–1148. [CrossRef]
8. Ramata-Stunda, A.; Valkovska, V.; Boroduškis, M.; Livkiša, D.; Kaktiņa, E.; Silamiķele, B.; Boroduške, A.; Pentjušs, A.; Rostoks, N. Development of metabolic engineering approaches to regulate the content of total phenolics, antiradical activity and organic acids in callus cultures of the highbush blueberry (*Vaccinium corymbosum* L.). *Agron. Res.* **2020**, *18*, 1860–1872. [CrossRef]
9. Pormale, J.; Osvalde, A.; Nollendorfs, V. Comparison study of cultivated highbush and wild blueberry nutrient status in producing plantings and woodlands. *Latv. J. Agron.* **2009**, *12*, 80–87.
10. Sas Paszt, L.; Sumorok, B.; Malusa, E.; Głuszek, S.; Derkowska, E. The influence of bioproducts on root growth and mycorrhizal occurrence in the rhizosphere of strawberry plants Elsanta. *J. Fruit Ornam. Plant Res.* **2011**, *19*, 13–34.
11. Koszański, Z.; Rumasz-Rudnicka, E.; Friedrich, S. Anatomy, morphology and yield of highbush blueberry (*Vaccinium corymbosum* L.) under the influence of irrigation and mineral fertilisation. *Acta Agrophys.* **2008**, *11*, 677–684.
12. Raj, A.B.; Raj, S.K. Zinc and boron nutrition in pulses: A review. *J. Appl. Nat. Sci.* **2019**, *11*, 673–679. [CrossRef]
13. Wójcik, P. Uptake of mineral nutrients from foliar fertilization. *J. Ornam. Plant Res.* **2004**, *12*, 201–218.
14. Jiang, Y.; Li, Y.; Zeng, Q.; Wei, J.; Yu, H. The effect of soil pH on plant growth, leaf chlorophyll fluorescence and mineral element content of two blueberries. *Acta Hortic.* **2017**, *1180*, 269–276. [CrossRef]
15. Kleiber, T. The Role of Silicon in Plant Tolerance to Abiotic Stress. In *Plant Nutrients and Abiotic Stress Tolerance*; Hasanuzzaman, M., Fujita, M., Oku, H., Nahar, K., Hawrylak-Nowak, B., Eds.; Springer: Singapore, 2018. [CrossRef]
16. Institution of Mechanical Engineers. Global Food. Waste Not, Want Not. Available online: http://www.imeche.org/docs/default-source/reports/Global_Food_Report.pdf?sfvrsn=0 (accessed on 15 May 2015).
17. Glonek, J.; Komosa, A. Fertigation of highbush blueberry (*Vaccinium corymbosum* L.). Part I. The effect on growth and yield. *Acta Sci. Pol. Hortorum Cultus* **2013**, *12*, 47–57.
18. Hiscox, J.D.; Israelstam, G.F. A method for the extraction of chlorophyll from leaf tissue without maceration. *Can. J. Bot.* **1978**, *5*, 1332–1334.
19. Bozan, B.; Başer, K.H.C.; Kara, S. Quantitative determination of naphthaquinones of Arnebia densiflora (Nordm.) Ledeb. by an improved high-performance liquid chromatographic method. *J. Chromatogr. A* **1997**, *782*, 133–136. [CrossRef]

20. Crisisto, C.H.; Slaughter, D.; Garner, D. Developing maximum maturity indices for "Tree ripe" fruit. *Adv. Hort. Sci.* **2004**, *18*, 29–32.
21. Spanos, G.A.; Wrolstad, R.E.; Heatherbell, D.A. Influence of processing and storage on the phenolic composition of apple juice. *J. Agric. Food Chem.* **1990**, *8*, 1572–1579. [CrossRef]
22. Wrolstad, R.E. Color and pigment analysis in fruit products. In *Station Bulletin*; University of Agricultural Experimental Station Oregon State: Corvallis, OR, USA, 1976; Volume 624, p. 16.
23. Dragišic Maksimović, J.; Bogdanovic, J.; Maksimović, V.; Nikolic, M. Silicon modulates the metabolism and utilization of phenolic compounds in cucumber (*Cucumis sativus* L.) grown atexcess manganese. *J. Plant Nutr. Soil Sci.* **2007**, *170*, 739–744. [CrossRef]
24. Silva Lobato, A.K.; Silva Guedes, E.M.; Marques, D.J.; de Oliveira Neto, C.F. Silicon: A benefic element to improve tolerance in plants exposed to water deficiency. In *Responses of Organisms to Water Stress*; Akıcnı, S., Ed.; InTech: Rijeka, Croatia, 2013.
25. Zydlik, Z.; Zydlik, P.; Kleiber, T. The effect of the mycorrhization on the content of macroelements in the soil and leaves of blueberry cultivated after replantation. *Zemdirbyste-Agriculture* **2019**, *106*, 345–350. [CrossRef]
26. Pliszka, K. Borówka wysoka czyli amerykańska. *Działkowiec* **2001**, *3*, 34–35.
27. Burkhard, N.; Lynch, D.; Percival, D.; Sharifi, M. Organic mulch impact on vegetation dynamics and productivity of highbush blueberry under organic production. *HortScience* **2009**, *44*, 688–696. [CrossRef]
28. Carroll, J.; Pritts, M.P.; Heidenreich, C. *Production Guide for Organic Blueberries*; N. Y. State Integrated Pest Manage Program: Ithaca, NY, USA, 2015.
29. Ochmian, I.; Kozos, K. Influence of foliar fertilisation with calcium fertilisers on the firmness and chemical composition of two highbush blueberry cultivars. *J. Elem.* **2015**, *20*, 185–201. [CrossRef]
30. Ochmian, I.; Oszmiański, J.; Jaśkiewicz, B.; Szczepanek, M. Fertilization of highbush blueberry with urea phosphate. *Folia Hortic.* **2018**, *30*, 295–305. [CrossRef]
31. Aliman, J.; Michalak, I.; Bušatlić, E.; Aliman, L.; Kulina, M.; Radovic, M.; Hasanbegovic, J. Study of the physicochemical properties of highbush blueberry and wild bilberry fruit in central Bosnia. *Turk. J. Agric. For.* **2020**, *44*, 156–168. [CrossRef]
32. Correia, S.; Gonçalves, B.; Aires, A.; Silva, A.; Ferreira, L.; Carvalho, R.; Fernandes, H.; Freitas, C.; Carnide, V.; Silva, A.P. Effect of harvest year and altitude on nutritional and biometric characteristics of blueberry cultivars. *J. Chem.* **2016**, *2016*, 8648609. [CrossRef]
33. Zorenc, Z.; Veberic, R.; Stampar, F.; Koron, D. Changes in berry quality of northern highbush blueberry (*Vaccinium corymbosum* L.) during the harvest season. *Turk. J. Agric. For.* **2016**, *40*, 855–864. [CrossRef]
34. Lanauskas, J.; Kviklienė, N.; Uselis, N.; Kviklys, D.; Buskienė, L.; Mažeika, R.; Staugaitis, G. The effect of calcium foliar fertilizers on cv. Ligol apples. *Plant Soil Environ.* **2012**, *58*, 466. [CrossRef]
35. Cortés-Rojas, M.E.; Mesa-Torres, P.A.; Grijalba-Rativa, C.M.; Pérez-Trujillo, M.M. Yield and fruit quality of the blueberry cultivars Biloxi and Sharpblue in Guasca, Colombia. *Agron. Colomb.* **2016**, *34*, 33–41. [CrossRef]
36. Shevchuk, L.M.; Grynyk, I.V.; Levchuk, L.M.; Yareshcenko, O.M.; Tereshcenko, Y.Y.; Babenko, S.M. Biochemical contents of highbush blueberry fruits grown in the Western Forest-Steppe of Ukraine. *Agron. Res.* **2021**, *19*, 232–249. [CrossRef]
37. Zenkova, M.; Pinchykova, J. Chemical composition of Sea-buckthorn and Highbush Blueberry fruits grown in the Republic of Belarus. *Food Sci. Appl. Biotechnol.* **2019**, *2*, 81–90. [CrossRef]
38. Celik, F.; Bozhuyuk, M.R.; Ercisli, S.; Gundogdu, M. Physicochemical and Bioactive Characteristics of Wild Grown Bilberry (*Vaccinium myrtillus* L.) Genotypes from Northeastern Turkey. *Not. Bot. Horti Agrobot. Cluj-Napoca* **2018**, *46*, 128–133. [CrossRef]
39. Ersoy, N.; Kupe, M.; Gundogdu, M.; Gulce, I.; Ercisli, S. Phytochemical and antioxidant diversity in fruits of currant (*Ribes* spp.) cultivars. *Not. Bot. Horti Agrobot. Cluj-Napoca* **2018**, *46*, 381–387. [CrossRef]
40. Ayour, J.; Sagar, M.; Alfeddy, M.; Taourirte, M.; Benichou, M. Evolution of pigments and their relationship with skin color based on ripening in fruits of different Moroccan genotypes of apricots (*Prunus armeniaca* L.). *Sci. Hort.* **2016**, *207*, 168–175. [CrossRef]
41. Echeverría, G.; Cantín, C.M.; Ortiz, A.; López, M.L.; Lara, I.; Graell, J. The impact of maturity, storage temperature and storage duration on sensory quality and consumer satisfaction of 'Big Top®' nectarines. *Sci. Hort.* **2015**, *190*, 179–186. [CrossRef]
42. Milukic-Petkovsek, M.; Ivancic, A.; Schimitzer, V.; Veberic, R.; Stampar, F. Comparison of major taste compounds and antioxidative properties of fruits and flowers of different Sambucus species and interspecific hybrids. *Food Chem.* **2016**, *200*, 134–140. [CrossRef]
43. Kim, H.-Y.; Farcuh, M.; Cohen, Y.; Crisosto, C.; Sadka, A.; Blumwald, E. Non-climacteric ripening and sorbitol homeosta-556 sis in plum fruits. *Plant Sci.* **2015**, *231*, 30–39. [CrossRef] [PubMed]
44. Li, X.; Li, C.; Sun, J.; Jackson, A. Dynamic changes of enzymes involved in sugar and organic acid level modification during blueberry fruit maturation. *Food Chem.* **2020**, *309*, 125617. [CrossRef]
45. Okan, O.T.; Deniz, I.; Yayli, N.; Şat, G.I.; Öz, M.; Serdar, G.H. Antioxidant activity, sugar content and phenolic profiling of blueberry cultivars: A comprehensive comparison. *Not. Bot. Horti Agrobot. Cluj-Napoca* **2018**, *46*, 639–652. [CrossRef]
46. Durán-Soria, S.; Pott, D.M.; Osorio, S.; Vallarino, J.G. Sugar Signaling During Fruit Ripening. *Front. Plant Sci.* **2020**, *11*, 564917. [CrossRef]
47. Fotirić Akšić, M.; Tosti, T.; Nedić, N.; Marković, M.; Ličina, V.; Milojković-Opsenica, D.; Tešić, Ž. Influence of frost damage on the sugars and sugar alcohol composition in quince (*Cydonia oblonga* Mill.) Floral nectar. *Acta Physiol. Plant.* **2015**, *37*, 1701. [CrossRef]
48. Kirina, I.B.; Belosokhov, F.G.; Titova, L.V.; Suraykina, I.A.; Pulpito, V.F. Biochemical assessment of berry crops as a source of production of functional food products. In *IOP Conference Series: Earth and Environmental Science*; IOP Publishing: Bristol, UK, 2020; Volume 548, p. 082068. [CrossRef]

49. Skrovankova, S.; Sumczynski, D.; Mlcek, J.; Jurikova, T.; Sochor, J. Bioactive compounds and antioxidant activity in different types of berries. *Int. J. Mol. Sci.* **2015**, *16*, 24673–24706. [CrossRef]

50. Prior, R.L.; Cao, G.A.; Martin, A. Antioxidant capacity as influenced by total phenolic and anthocyanin content, maturity, and variety of vaccinium species. *J. Agric. Food Chem.* **1998**, *46*, 2686–2693. [CrossRef]

51. Rupasova, Z.; Pavlovskij, N.; Kurlovich, T.; Pyatnitsa, F.; Yakovlev, A.; Volotovich, A.; Pinchukova, Y. Variability of structure of biochemical composition of fruits of the highbush blueberry. *Agron. Vestis Latv. J. Agron.* **2009**, *12*, 103–107.

52. Giacalone, M.; Di Sacco, F.; Traupe, I.; Pagnucci, N.; Forfori, F.; Giunta, F. Blueberry polyphenols and neuroprotection. In *Bioactive Nutraceuticals and Dietary Supplements in Neurological and Brain Disease, Prevention and Therapy*, 1st ed.; Watson, R.R., Preedy, V.R., Eds.; Academic Press: Cambridge, MA, USA, 2015; pp. 17–28. [CrossRef]

53. Sadowska, A.; Dybkowska, E.; Rakowska, R.; Hallmann, E.; Świderski, F. Ocena zawartości składników bioaktywnych i właściwości przeciwutleniających proszków wyprodukowanych metodą liofilizacji z wybranych surowców roślinnych. *Żywność Nauka Technol. Jakość* **2017**, *24*, 59–75. [CrossRef]

54. Gramza-Michałowska, A.; Człapka-Matyasik, M. Evaluation of the antiradical potential of fruit and vegetable snacks. *Acta Sci. Pol. Technol. Aliment.* **2011**, *10*, 61–72.

55. Gao, L.; Mazza, G. Quantitation and distribution of simple and acylated anthocyanins and other phenolics in blueberries. *J. Food Sci.* **1994**, *59*, 1057–1059. [CrossRef]

56. Starast, M.; Karp, K.; Vool, E.; Moor, U.; Tonutare, T.; Paal, T. Chemical composition and quality of cultivated and natural blueberry fruit in Estonia. *Veg. Crop. Res. Bull.* **2007**, *66*, 143–153. [CrossRef]

57. Sams, C.E. Preharvest factors affecting postharvest texture. *Postharvest Biol. Technol.* **1999**, *15*, 249–254. [CrossRef]

horticulturae

Article

Quality Assessment of Dried White Mulberry (*Morus alba L.*) Using Machine Vision

Adel Hosainpour [1,*], Kamran Kheiralipour [2], Mohammad Nadimi [3] and Jitendra Paliwal [3]

[1] Mechanical Engineering of Biosystems Department, Urmia University, Urmia 5756151818, Iran
[2] Mechanical Engineering of Biosystems Department, Ilam University, Ilam 6939177111, Iran
[3] Department of Biosystems Engineering, University of Manitoba, Winnipeg, MB R3T 5V6, Canada
* Correspondence: a.hosainpour@urmia.ac.ir; Tel./Fax: +98-4432775035

Abstract: Over the past decade, the fresh white mulberry (*Morus alba L.*) fruit has gained growing interest due to its superior health and nutritional characteristics. While white mulberry is consumed as fresh fruit in several countries, it is also popular in dried form as a healthy snack food. One of the main challenges that have prevented a wider consumer uptake of this nutritious fruit is the non-uniformity in its quality grading. Therefore, identifying a reliable quality grading tool can greatly benefit the relevant stakeholders. The present research addresses this need by developing a novel machine vision system that combines the key strengths of image processing and artificial intelligence. Two grades (i.e., high- and low-quality) of white mulberry were imaged using a digital camera and 285 colour and textural features were extracted from their RGB images. Using the quadratic sequential feature selection method, a subset of 23 optimum features was identified to classify samples into two grades using artificial neural networks (ANN) and support vector machine (SVM) classifiers. The developed system under both classifiers achieved the highest correct classification rate (CCR) of 100%. Indeed, the latter approach offered a smaller mean squared error for the training and test sets. The developed model's high accuracy confirms the machine vision's suitability as a reliable, low-cost, rapid, and intelligent tool for quality monitoring of dried white mulberry.

Keywords: dried mulberry; quality grading; image processing; feature classification; artificial neural networks

Citation: Hosainpour, A.; Kheiralipour, K.; Nadimi, M.; Paliwal, J. Quality Assessment of Dried White Mulberry (*Morus alba L.*) Using Machine Vision. *Horticulturae* **2022**, *8*, 1011. https://doi.org/10.3390/horticulturae8111011

Academic Editor: Esmaeil Fallahi

Received: 27 September 2022
Accepted: 27 October 2022
Published: 1 November 2022

Publisher's Note: MDPI stays neutral with regard to jurisdictional claims in published maps and institutional affiliations.

1. Introduction

With an increasing demand for healthy and nutritious agri-food products, berries have gained a lot of interest around the globe. For instance, the global import value of fresh mulberries, raspberries, loganberries, and blackberries has risen from USD 1.7 billion in 2014 to USD 3.8 billion in 2021. The top five importers of berries above include the United States, Germany, Canada, United Kingdom, and Spain, with 43%, 10%, 9%, 9%, and 6% of the global market import, respectively [1].

Among berry fruits, white mulberry is of great interest as it contains carbohydrates, protein, fiber, fat, vitamins and minerals [2,3]. Moreover, the phenolic compounds of white mulberry have a wide range of antioxidant and antimutagenic activities and anti-cancer properties [3–5].

White mulberry can be consumed both in fresh or dried form (as a healthy snack food) [4]. However, the main challenge the industry is facing in trading this nutritious fruit, specifically in its dried form, is the inconsistency in its quality grading due to a lack of well-developed tools. The quality of the fresh fruit and the drying conditions are the main factors that determine the dried white mulberry quality. High-quality products usually have negligible damaged/broken components and are milky in colour. In contrast, lower-quality dried samples have several damaged/broken/smashed pieces and are darker in colour. The industry's common approach for grading dried mulberry is a visual inspection conducted by skilled personnel. However, manual assessment is time-consuming, expensive, and a subjective task. Therefore, developing a rapid and intelligent grading tool can benefit the industry and provide consistent quality to consumers.

Quality monitoring is an important step in production to achieve high-marketability products with optimum use of resources [6]. Over the past decade, the potential of optical and machine vision techniques for agri-food products' quality monitoring has remarkably progressed [7–14]. For instance, machine vision systems have been utilized for the evaluation of agri-food products' maturity levels [15,16], mechanical damage [17–19], soundness [20–22], growth parameters [18,23], and grading [24–26].

The most frequently used machine vision tools have been imaging- or spectroscopic-based systems [27,28]. The former offers lower costs, while the latter has the capability to explore a sample's chemical constituents. Indeed, the majority of developed models have been sample- and application-specific due to the nature of the products. So, each product and/or application needs its own calibration and algorithm tools [16].

Scholars have demonstrated the application of imaging-based machine vision tools for the maturity evaluation of various products, including fresh white mulberry [29]. However, despite several agri-food products being extensively researched for non-destructive quality determination, our thorough literature review indicates that there has been no attempt to explore machine vision for the quality evaluation of dried mulberry. The present work aims to fill this knowledge gap by developing a novel machine vision system that combines the key strengths of image processing and artificial intelligence. The authors believe the outcome of the present study can open new horizons toward automating mulberry quality monitoring.

2. Materials and Methods

2.1. Samples

The dried white mulberry (*Morus alba L.*) (Figure 1a) product was purchased from a local market in Marivan City, Kurdistan, Iran. A panel of experts manually graded the samples into two grades (i.e., high- and low-quality) (Figure 1b) based on the appearance of the product. A total of 100 representative samples were selected for this study (50 samples for each grade).

(a)   (b)   (c)

Figure 1. High- and low-quality dried white mulberry samples, (**a**) mixed combination, (**b**) manual grouping, and (**c**) magnified view of an individual sample.

2.2. Image Acquisition

The individual samples were placed on white paper to be imaged (see Figure 1c) using a digital camera (model J7, Samsung Corp., Seoul, Korea). The camera's resolution was 13 MP.

2.3. Image Processing

The image processing steps involved preprocessing, feature extraction and feature analysis. Figure 2 depicts the proposed algorithm's different steps, which will be discussed in further detail later. The image analysis was performed in MATLAB (Version 2016a, Mathworks Inc., Waltham, MA, USA).

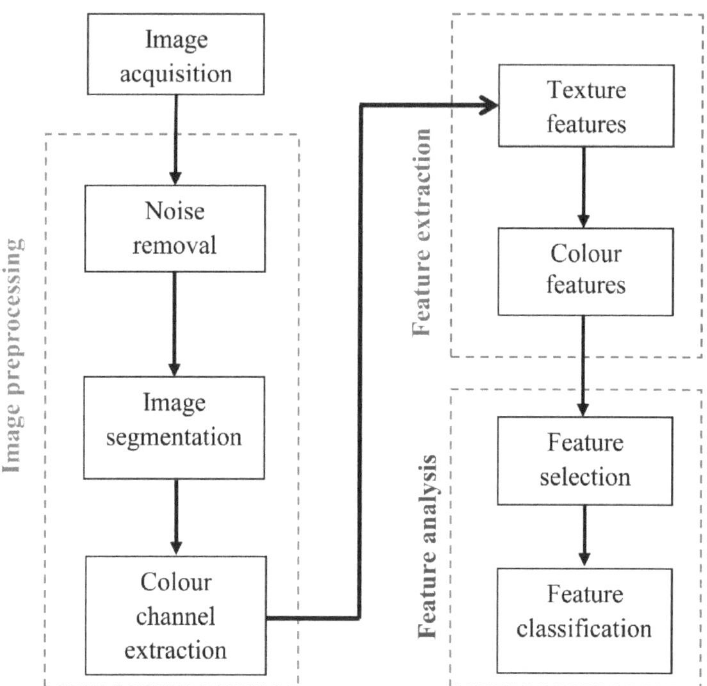

Figure 2. The flowchart of the machine vision system.

2.3.1. Image Preprocessing

To separate the object's image from the background, the acquired RGB (Red Green Blue) images were processed using Equation (1) (empirical equation):

$$BI = R - 0.5G - 0.5B \tag{1}$$

where BI is a monocolour image [30,31], and R, G, and B are the channels of the sample image in RGB colour space, respectively. The obtained BI images were converted to binary (black and white) images to form binary masks. Erosion and dilation operations were employed for noise removal. The initial RGB images were multiplied by the binary masks to select regions of interest (ROI).

2.3.2. Feature Extraction

The successful use of various colour and texture features in agri-food quality evaluation has been previously demonstrated [29,32]. Herein, a similar approach to [16] was utilized. First, the images in RGB colour space were transferred to six other colour spaces, including L*a*b*, HSV, NRGB, CrCgCb, I1I2I3, and gray level spaces [25,30,33,34]. Detailed information about the provided colour spaces and the corresponding individual channels are available elsewhere [30,35–37]. The data of utilized colour spaces were recorded from 19 channels, including R, G, B, L*, a*, b*, H, S, V, nr, ng, nb, cr, cg, cb, I1, I2, I3, and gray level [16].

From the colour spaces mentioned above, 10 colour features were calculated, namely maximum, mean, minimum, mode, median, standard deviation, coefficient of variation, kurtosis, skewness, and covariance [15,31,35,38]. Moreover, from the gray-level co-occurrence matrix (GLCM) of ROI images, 5 texture features were calculated, namely energy, entropy, correlation, homogeneity and contrast [25,30,34]. Overall, for each sample, 285 features were recorded (15 features × 19 channels).

2.3.3. Feature Analysis
2.3.3.1. Feature Selection

It is well-established that when dealing with large data sets, feature selection could reduce computational time and power while enhancing efficiency. This work employed a quadratic-based sequential feature selection method according to [16,30,31,34,39] to identify optimum features among the available 285 features (see Section 2.3.2).

2.3.3.2. Feature Classification

Shallow artificial neural networks (ANNs) [40–43] and support vector machine (SVM) methods [38] were used to classify the fruit into two classes: high- and low-quality.

The ANN structure consisted of three (viz. input, hidden and target) layers. The number of optimum features (Section 2.3.3.1) and desired classes (i.e., high- and low-quality mulberry) determined the number of neurons in the input and target layer, respectively. The number of neurons in the hidden layer varied from 2 to 20 to identify the optimum structure. The activation functions for the hidden and target layers were *tangent sigmoid* and *pure line*, respectively [16]. The data were randomly divided into three groups: 60% for training, 20% for validation, and 20% for testing. Two main measures were evaluated to assess the performance of various ANN-based classifiers: the correlation coefficient (R) and correct classification rate (CCR).

In the case of the SVM classifier, the data were divided into two groups: 75% for training and 25% for testing. The results of the SVM method were compared with that of ANN in terms of CCR and mean squared error (MSE).

3. Results
3.1. Feature Set Optimization

Using the feature selection algorithm (Section 2.3.3.1), 23 out of 285 features were identified as optimum. As seen in Table 1, the mean values of the optimum features were substantially different for high- and low-quality mulberry. These optimum features were used as the inputs of the ANN classifiers.

Table 1. The optimum features of dried white mulberry. Grades 1 and 2 represent high- and low-quality samples, respectively.

No	Feature	Channel	Mean	
			Grade 1	Grade 2
1	Mean	Gray	0.48	0.35
2	Median	Gray	0.48	0.34
3	Mean	R	0.52	0.38
4	Coefficient of variation	R	0.08	0.14
5	Median	R	0.52	0.38
6	Mean	G	0.47	0.33
7	Median	G	0.47	0.33

Table 1. *Cont.*

No	Feature	Channel	Mean	
			Grade 1	**Grade 2**
8	Mean	B	0.42	0.31
9	Kurtosis	B	3.68	5.58
10	Mode	B	0.417	0.30
11	Mean	L*	74.44	65.19
12	Standard deviation	L*	2.80	3.82
13	Median	L*	74.75	64.88
14	Skewness	L*	3.23	4.85
15	Coefficient of variation	L*	8.07	15.00
16	Mode	I1	0.48	0.32
17	Entropy	I3	1.30	0.19
18	Median	H	0.09	0.07
19	Mean	V	0.52	0.38
20	Standard deviation	V	0.04	0.05
21	Coefficient of variation	V	0.08	0.14
22	Median	V	0.52	0.38
23	Mode	V	0.53	0.37

3.2. Feature Classification

Table 2 shows the performance of different ANN structures discussed in Section 2.3.3.2. The best performance was achieved under the 23-14-2 structure (the three numbers represent the neurons in the input, hidden, and output layers, respectively). The CCR for training, validation, testing and entire data set under the 23-14-2 structure was 100%. The correlation coefficient for the training, validation and the whole data set was 1.

Table 2. The developed ANN classification structures with their characteristics.

No.	Structure	Training Data		Validation Data		Testing Data		Total Data	
		R *	**CCR ****	**R**	**CCR**	**R**	**CCR**	**R**	**CCR**
1	23-3-2	1.00	100.00	1.00	100.00	0.90	95.00	0.98	99.00
2	23-4-2	1.00	100.00	1.00	100.00	0.90	95.00	0.98	99.00
3	23-5-2	1.00	100.00	1.00	100.00	0.90	95.00	0.98	99.00
4	23-6-2	1.00	100.00	1.00	100.00	0.90	95.00	0.98	99.00
5	23-7-2	1.00	100.00	1.00	100.00	0.90	95.00	0.98	99.00
6	23-8-2	1.00	100.00	1.00	100.00	0.90	95.00	0.98	99.00
7	23-9-2	1.00	100.00	1.00	100.00	0.90	95.00	0.98	99.00
8	23-10-2	1.00	100.00	1.00	100.00	0.92	95.00	0.98	99.00
9	23-11-2	1.00	100.00	0.99	100.00	0.69	95.00	0.93	99.00
10	23-12-2	1.00	100.00	1.00	100.00	0.74	95.00	0.94	99.00
11	23-13-2	1.00	100.00	1.00	100.00	0.86	95.00	0.97	99.00
12	23-14-2	1.00	100.00	1.00	100.00	0.99	100.00	1.00	100
13	23-15-2	1.00	100.00	1.00	100.00	0.87	95.00	0.97	99.00
14	23-16-2	1.00	100.00	1.00	100.00	0.81	100.00	0.96	100.00
15	23-17-2	1.00	100.00	1.00	100.00	0.87	95.00	0.97	99.00
16	23-18-2	1.00	100.00	0.92	90.00	0.66	95.00	0.90	97.00
17	23-19-2	1.00	100.00	1.00	100.00	0.84	95.00	0.97	99.00
18	23-20-2	1.00	100.00	0.99	100.00	0.86	95.00	0.97	99.00

* R represents the correlation coefficient. ** CCR represents the correct classification rate (%).

The confusion matrices of the optimum model for training, validation, testing, and the whole data set are presented in Table 3. The optimum classifier can successfully discriminate different grades of the dried white mulberry in all data sets.

Table 3. The confusion matrix of the optimum classifier (ANN with 23-14-2 structure). Grades 1 and 2 represent the number of high- and low-quality samples, respectively.

Data Set	Actual	Predicted		CCR
		Grade 1	Grade 2	
Training	Grade 1	30	0	100%
	Grade 2	0	30	
Validation	Grade 1	10	0	100%
	Grade 2	0	10	
Testing	Grade 1	10	0	100%
	Grade 2	0	10	
Total	Grade 1	50	0	100%
	Grade 2	0	50	

Figure 3 shows the MSE of the optimum classifier model in the validation step. This figure gave the minimum value of the mean squared error as 0.00066 after six epochs. The MSEs for training and test set at the same epoch were around 0.000005 and 0.01, respectively. Figure 4 illustrates the regression lines of the optimum classifier model for training, validation, testing, and the whole data set. The correlation coefficients of the classifier for all datasets were in excess of 0.996, except for the test data set (0.985).

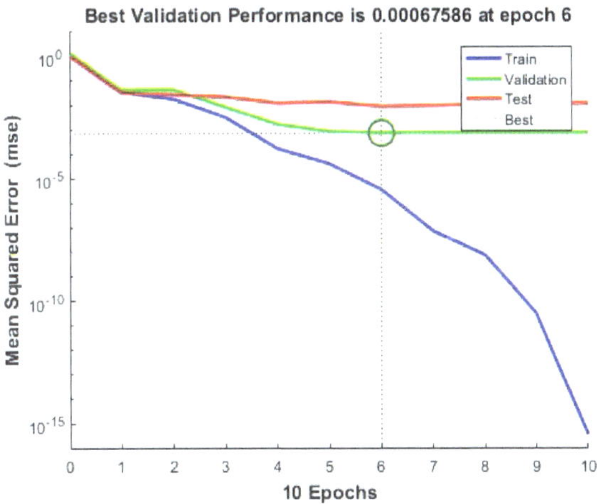

Figure 3. The mean squared error (MSE) of the optimum classifier model.

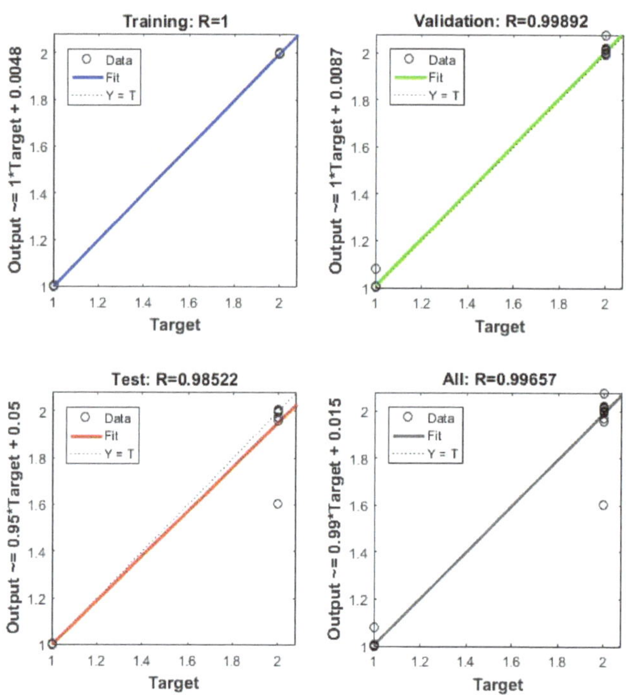

Figure 4. The regression lines of the optimum classifier model. 1 and 2 refer to groups 1 and 2, respectively.

The results of the SVM classifier for training, testing, and the whole data set have been presented in Table 4. It is apparent that the two grades can be distinguished by 100 % accuracy. The MSE values of the classification model for training, testing, and the whole data set were 0.013, 0.040, and 0.01, respectively.

Table 4. The confusion matrix of the SVM classifier. Grades 1 and 2 represent the number of high- and low-quality samples, respectively.

Data Set	Actual	Predicted		CCR
		Grade 1	Grade 2	
Training	Grade 1	38	0	100%
	Grade 2	0	37	
Testing	Grade 1	12	0	100%
	Grade 2	0	13	
Total	Grade 1	50	0	100%
	Grade 2	0	50	

4. Discussion

Our results confirm the imaging-based machine vision tool's capability for rapid quality monitoring of dried white mulberry. Both SVM and ANN classifiers provided 100% accuracy in distinguishing the high- and low-quality dried mulberry. Indeed, the latter approach can be considered superior as it offers lower MSE.

To the best of our knowledge, the present work is the first effort at quality monitoring of dried white mulberry using a low-cost machine vision system. The obtained results are in reasonable agreement with similar maturity classifiers reported by Kheiralipour et al. [16] and Salam et al. [31]. We believe the proposed system is useful for improving product marketability by sending high-quality products with uniform colour and texture to the target markets. Moreover, the system is beneficial from a waste management perspective by sending the lower-quality products to special processing units, which can be used as input for different products, such as pet foods. Hence, this research can be very useful in enhancing the postharvest quality determination and processing of the white mulberry fruit.

To improve the reliability of the developed model, future works need to evaluate the performance of the proposed intelligent algorithms on larger datasets. Acquiring more datasets may be achieved by imaging more samples or through implementing data augmentation tools such as Generative Adversarial Networks (GANs) [44].

To facilitate the uptake of the developed machine vision tool by farmers and industry stakeholders, the system should segregate the samples according to their quality parameters in a rapid and real-time manner. One strategy to reach such a goal is to integrate the imaging tools into an intelligent conveyer belt-based classifier system, similar to the one proposed by Azarmdel et al. [29]. Hence, another potential topic for future research could be to evaluate the performance of the proposed machine vision tools on such an opto-electro-mechanical system [29]. Once a reliable performance is achieved, the system can be integrated into the mulberry drying industry for quality control and optimization.

One should note that the capital costs associated with the proposed quality assessment system may be the main prohibiting factor for the farmers and industry to uptake the mentioned technology. However, the return on such extra capital cost, especially for commercial installations will be faster processing, improved classification accuracy, and the need for fewer human personnel. Thus, it can be considered a safe, short-run return and reasonable investment, which may even qualify for government subsidies in some jurisdictions.

Ultimately, one should consider that while the developed image processing algorithm in the present research are sample-specific, they can be modified to apply to other similar fruit such as figs (*Ficus carica*) and other berries. Moreover, upon further modifications and optimization, the proposed model may be able to classify fruits according to other quality characteristics such as soundness and maturity levels.

5. Conclusions

A rapid, low-cost, non-destructive machine vision-based approach was developed to assess the quality of dried white mulberry samples. Using SVM and ANN classifiers, the samples were classified into two grades (i.e., high- and low-quality) with 100% correct classification accuracy. The mean squared error of the test set for the aforementioned classifiers were 0.04 and 0.01, respectively, indicating the better performance of the ANN classifier.

Considering the promising performance of the proposed machine vision system, the future effort should focus on adopting the proposed technology to develop an opto-electro-mechanical system that can operate in real-time and at an industrial scale. Upon modification and optimization, such systems will enable rapid and automated grading of mulberry (or similar fruits) based on various characteristics such as appearance, soundness and maturity levels.

Author Contributions: Conceptualization, A.H. and K.K.; methodology, A.H. and K.K.; software, K.K.; formal analysis, K.K. and M.N.; investigation A.H. and K.K.; resources, A.H. and K.K.; data curation, A.H. and K.K.; writing-original draft preparation, A.H. and K.K.; writing—review and editing, M.N., J.P.; project administration, A.H., K.K., M.N., J.P.; funding acquisition, A.H., K.K., M.N., and J.P. All authors have read and agreed to the published version of the manuscript.

Funding: This research was funded by the Natural Sciences and Engineering Research Council of Canada (NSERC) grant number RGPIN-2021-03350. The authors also would like to thank the financial support provided by Ilam University, and Urmia University.

Data Availability Statement: The data that support the findings of this study are available from the corresponding authors, upon reasonable request.

Conflicts of Interest: The authors declare that they have no known competing financial interest or personal relationships that could have appeared to influence the work reported in this paper.

References

1. Tridge. Available online: https://www.tridge.com/trades/081020-raspberry-blackberry-mulberry-and-loganberry-fresh/import (accessed on 20 September 2022).
2. Singhal, B.K.; Khan, M.A.; Dhar, A.; Baqual, F.M.; Bindroo, B.B. Approaches to industrial exploitation of mulberry (Mulberry sp.) fruits. *J. Fruit Ornam. Plant Res.* **2010**, *18*, 83–99.
3. Malgorzata, L. Energy and nutritional properties of the white mulberry (*Morus alba* L.). *J. Agric. Sci. Technol.* **2015**, *A5*, 709–716.
4. Elhamirad, A.H. Optimization of juice clarification of two native white mulberry (*Morus alba*) varieties. *Iran. J. Innov. Food Sci. Technol.* **2013**, *5*, 91–103.
5. Yuan, Q.; Zhao, L. The Mulberry (*Morus alba* L.) Fruit-A Review of Characteristic Components and Health Benefits. *J. Agric. Food Chem.* **2017**, *65*, 10383–10394. [CrossRef] [PubMed]
6. Kheiralipour, K. *Sustainable Production: Definitions, Aspects, and Elements*, 1st ed.; Nova Science Publisher: Hauppauge, NY, USA, 2022.
7. Erkinbaev, C.; Nadimi, M.; Paliwal, J. A unified heuristic approach to simultaneously detect fusarium and ergot damage in wheat. *Meas. Food* **2022**, *7*, 100043. [CrossRef]
8. Karunakaran, C.; Visen, N.S.; Paliwal, J.; Zhang, G.; Jayas, D.S.; White, N.D.G. Machine Vision Systems for Agricultural Products. In *Introduction to Advanced Food Process Engineering*; Sahu, J.K., Boca Raton, F.L., Eds.; Taylor & Francis Group: Oxfordshire, England, 2014.
9. Kumar Mahanti, N.; Pandiselvam, R.; Kothakota, A.; Ishwarya, P.S.; Kumar Chakraborty, S.; Kumar, M.; Cozzolino, D. Emerging non-destructive imaging techniques for fruit damage detection: Image processing and analysis. *Trends Food Sci. Technol.* **2022**, *120*, 418–438. [CrossRef]
10. Li, X.; Guillermic, R.M.; Nadimi, M.; Paliwal, J.; Koksel, F. Physical and microstructural quality of extruded snacks made from blends of barley and green lentil flours. *Cereal Chem.* **2022**, *99*, 1112–1123. [CrossRef]
11. Nadimi, M.; Sun, D.W.; Paliwal, J. Recent applications of novel laser techniques for enhancing agricultural production. *Laser Phys.* **2021**, *31*, 053001. [CrossRef]
12. Pathmanaban, P.; Gnanavel, B.K. Sundaram Anandan, S.Recent application of imaging techniques for fruit quality assessment. *Trends Food Sci. Technol.* **2019**, *94*, 32–42. [CrossRef]
13. Rafiq, A.; Makroo, H.A.; Hazarika, M.K. Artificial Neural Network-Based Image Analysis for Evaluation of Quality Attributes of Agricultural Produce. *J. Food Process. Preserv.* **2016**, *40*, 1010–1019. [CrossRef]
14. Sivakumar, C.; Chaudhry, M.M.A.; Nadimi, M.; Paliwal, J.; Courcelles, J. Characterization of roller and Ferkar-milled pulse flours using laser diffraction and scanning electron microscopy. *Powder Technol.* **2022**, *409*, 117803. [CrossRef]
15. Mohammadi, V.; Kheiralipour, K.; Ghasemi Varnamkhasti, M. Detecting maturity of persimmon fruit based on image processing technique. *Sci. Hortic.* **2015**, *184*, 123–128. [CrossRef]
16. Kheiralipour, K.; Nadimi, M.; Paliwal, J. Development of an Intelligent Imaging System for Ripeness Determination of Wild Pistachios. *Sensors* **2022**, *22*, 7134. [CrossRef] [PubMed]
17. Nadimi, M.; Loewen, G.; Paliwal, J. Assessment of mechanical damage to flaxseeds using radiographic imaging and tomography. *Smart Agric. Technol.* **2022**, *2*, 100057. [CrossRef]
18. Nadimi, M.; Sun, D.W.; Paliwal, J. Effect of laser biostimulation on germination of wheat. *Appl. Eng. Agric.* **2022**, *38*, 77–84. [CrossRef]
19. Nadimi, M.; Divyanth, L.G.; Paliwal, J. Automated detection of mechanical damage in flaxseeds using radiographic imaging and machine learning. *Foods Bioprocess Technol.* 2022; *In Press*.
20. Kheiralipour, K.; Ahmadi, H.; Rajabipour, A.; Rafiee Javan-Nikkhah MJayas, D.S.; Siliveu, K. Detection of fungal infection in pistachio kernel by long-wave near-infrared hyperspectral imaging technique. *Qual. Assur. Saf. Crops Foods* **2015**, *8*, 129–135. [CrossRef]
21. Kheiralipour, K.; Ahmadi, H.; Rajabipour, A.; Rafiee, S.; Javan-Nikkhah, M.; Jayas, D.S.; Siliveu, K.; Malihipour, A. Processing the hyperspectral images for detecting infection of pistachio kernel by R5 and KK11 isolates of Aspergillus flavus fungus. *Iran. J. Biosyst. Eng.* **2021**, *52*, 13–25.
22. Nadimi, M.; Brown, J.M.; Morrison, J.; Paliwal, J. Examination of wheat kernels for the presence of Fusarium damage and mycotoxins using near-infrared hyperspectral imaging. *Meas. Food* **2021**, *4*, 100011. [CrossRef]
23. Nadimi, M.; Loewen, G.; Bhowmik, P.; Paliwal, J. Effect of laser biostimulation on germination of sub-optimally stored flaxseeds (*Linum usitatissimum*). *Sustainability* **2022**, *14*, 12183. [CrossRef]
24. Kheiralipour, K.; Pormah, A. Introducing new shape features for classification of cucumber fruit based on image processing technique and artificial neural networks. *J. Food Process Eng.* **2017**, *40*, e12558. [CrossRef]
25. Jahanbakhshi, A.; Kheiralipour, K. Evaluation of image processing technique and discriminant analysis methods in postharvest processing of carrot fruit. *Food Sci. Nutr.* **2020**, *8*, 3346–3352. [CrossRef] [PubMed]
26. Kheiralipour, K.; Kazemi, A. A new method to determine morphological properties of fruits and vegetables by image processing technique and nonlinear multivariate modeling. *Int. J. Food Prop.* **2020**, *23*, 368–374. [CrossRef]
27. Vadivambal, R.; Jayas, D.S. *Bio-Imaging: Principles, Techniques, and Applications*, 1st ed.; CRC Press: Boca Raton, FL, USA, 2018.

28. Kheiralipour, K.; Ahmadi, H.; Rajabipour, A.; Rafiee, S. *Thermal Imaging, Principles, Methods and Applications*, 1st ed.; Ilam University Publication: Ilam, Iran, 2018.
29. Azarmdel, H.; Jahanbakhshi, A.; Mohtasebi, S.S.; Rosado Muñoz, A. Evaluation of image processing technique as an expert system in mulberry fruit grading based on ripeness level using artificial neural networks (ANNs) and support vector machine (SVM). *Postharvest Biol. Technol.* **2020**, *166*, 111201. [CrossRef]
30. Khazaee, Y.; Kheiralipour, K.; Hosainpour, A.; Javadikia, H.; Paliwal, J. Development of a Novel Image Analysis and Classification Algorithms to Separate Tubers from Clods and Stones. *Potato Res.* **2022**, *65*, 1–22. [CrossRef]
31. Salam, S.; Kheiralipour, K.; Jian, J. Detection of Unripe Kernels and Foreign Materials in Chickpea Mixtures Using Image Processing. *Agriculture* **2022**, *12*, 995. [CrossRef]
32. Sabzi, S.; Nadimi, M.; Abbaspour-Gilandeh, Y.; Paliwal, J. Non-Destructive Estimation of Physicochemical Properties and Detection of Ripeness Level of Apples Using Machine Vision. *Int. J. Fruit Sci.* **2022**, *22*, 628–645. [CrossRef]
33. Gonzalez, R.; Woods, R. *Digital Image Processing*, 3rd ed.; Prentice-Hall: Hoboken, NJ, USA, 2007.
34. Azadnia, R.; Kheiralipour, K. Evaluation of hawthorns maturity level by developing an automated machine learning-based algorithm. *Ecol. Inform.* **2022**, *71*, 101804. [CrossRef]
35. Azadnia, R.; Kheiralipour, K. Recognition of leaves of different medicinal plant species using a robust image processing algorithm and artificial neural networks classifier. *J. Appl. Res. Med. Aromat. Plants* **2021**, *25*, 100327. [CrossRef]
36. García-Mateos, G.; Hernández-Hernández, J.L.; Escarabajal-Henarejos, D.; Jaén-Terrones, S.; Molina-Martínez, J.M. Study and comparison of color models for automatic image analysis in irrigation management applications. *Agric. Water Manag.* **2015**, *151*, 158–166. [CrossRef]
37. Chaves-González, J.M.; Vega-Rodríguez, M.A.; Gómez-Pulido, J.A.; Sánchez-Pérez, J.M. Detecting skin in face recognition systems: A colour spaces study. *Digit. Signal Process.* **2010**, *20*, 806–823. [CrossRef]
38. Jahanbakhshi, A.; Kheiralipour, K. Carrot sorting based on shape using image processing, artificial neural network, and support vector machine. *J. Agric. Mach.* **2019**, *9*, 295–307.
39. Available online: https://www.mathworks.com/help/stats/sequential-feature-selection.html (accessed on 14 October 2022).
40. Kheiralipour, K. Implementation and Construction of a System for Detecting Fungal Infection of Pistachio Kernel Based on Thermal Imaging (TI) and Image Processing Technology. Ph.D. Dissertation, University of Tehran, Karaj, Iran, 2012.
41. Williams, H.A.M.; Jones, M.H.; Nejati, M.; Seabright, M.J.; Bell, J.; Penhall, N.D.; Barnett, J.J.; Duke, M.D.; Scarfe, A.J.; Ahn, H.S.; et al. Robotic kiwifruit harvesting using machine vision, convolutional neural networks, and robotic arms. *Biosyst. Eng.* **2019**, *181*, 140–156. [CrossRef]
42. Kheiralipour, K.; Ahmadi, H.; Rajabipour, A.; Rafiee, S.; Javan-Nikkhah, M. Classifying healthy and fungal infected-pistachio kernel by thermal imaging technology. *Int. J. Food Prop.* **2015**, *18*, 93–99. [CrossRef]
43. Kheiralipour, K.; Marzbani, F. Pomegranate quality sorting by image processing and artificial neural network. In Proceedings of the 10th Iranian National Congress on AGR Machi Eng (Biosystems) and Mechanizasion, Mashhad, Iran, 30–31 August 2016. (In Persian).
44. Divyanth, L.G.; Guru, D.S.; Soni, P.; Machavaram, R.; Nadimi, M.; Paliwal, J. Image-to-image translation-based data augmentation for improving crop/weed classification models for precision agriculture applications. *Algorithms*, 2022; *in press*. [CrossRef]

 *horticulturae*

Article

An Economic Comparison of High Tunnel and Open-Field Strawberry Production in Southeastern Virginia

Jean Claude Mbarushimana [1], Darrell J. Bosch [1,*] and Jayesh B. Samtani [2]

1 Department of Agricultural and Applied Economics, Virginia Tech, Blacksburg, VA 24060, USA
2 Hampton Roads Agricultural Research and Extension Center, School of Plant and Environmental Sciences, Virginia Tech, Virginia Beach, VA 23455, USA
* Correspondence: bosch@vt.edu; Tel.: +1-540-231-5265

Abstract: High tunnels have been reported to extend the harvest season for fruits and vegetables in several North American regions. This study was conducted to evaluate whether there are additional economic returns from strawberries produced in high tunnel structures compared to open-field in the Commonwealth of Virginia. A total of eight strawberry cultivars were evaluated in a randomized complete block under high tunnel and open-field conditions. Total costs were estimated for all eight cultivars under high tunnel and open-field, and gross and net revenues from all cultivars were estimated over three marketing strategies (pre-pick wholesale, pre-pick retail, and U-pick) for both high tunnel and open-field. The average net revenues per hectare in the high tunnel were −\$62,077 (−\$25,122 ac^{-1}), −\$15,151 (−\$6131 ac^{-1}), and −\$27,938 (−\$11,306 ac^{-1}) for pre-pick wholesale, pre-pick retail, and U-pick, respectively, compared to open-field net revenues of \$39,816 (\$16,113 ac^{-1}), \$112,102 (45,366 ac^{-1}), and \$81,850 (\$33,123 ac^{-1}) for wholesale, pre-pick retail, and U-pick, respectively. Net revenues in the high tunnel were lower due to lower yields and higher production costs including overhead cost of the high tunnel structure. Almost all cultivars in the high tunnel generated negative net revenues regardless of the marketing strategy. The exceptions were 'Camino Real' which generated positive net revenues with U-pick and pre-pick retail marketing and 'Merced' which generated positive net revenues for pre-pick retail marketing. In contrast, net revenues from open-field cultivars were always positive. Results imply that growers should focus on open-field rather than high-tunnel strawberry production. Results are from one season of production. Replication of the study under one or more production seasons would contribute to more robust findings of the economic viability of strawberry production under a high tunnel.

Keywords: annual hill plasticulture; cultivar; cost; revenue; strawberry; net revenue; pre-pick; wholesale; retail; U-pick; economic comparison

Citation: Mbarushimana, J.C.; Bosch, D.J.; Samtani, J.B. An Economic Comparison of High Tunnel and Open-Field Strawberry Production in Southeastern Virginia. *Horticulturae* **2022**, *8*, 1139. https://doi.org/10.3390/horticulturae8121139

Academic Editors: Christian Fischer and Francesco Sottile

Received: 30 August 2022
Accepted: 24 November 2022
Published: 2 December 2022

Publisher's Note: MDPI stays neutral with regard to jurisdictional claims in published maps and institutional affiliations.

1. Introduction

In 2018, 19,919 hectares (49,220 ac) of strawberries were harvested in the United States of America. California harvested 72.7% of the total area with 14,488 hectares (35,800 ac), followed by Florida (19.9%). Oregon, Washington, North Carolina, and New York accounted for approximately 7.4% of total strawberry production with 1465 hectares (3620 ac) [1]. South Atlantic states (Alabama, Georgia, North Carolina, South Carolina, and Virginia) collectively produce strawberries on 948 hectares (2342 ac) [2]. In the Commonwealth of Virginia, strawberries occupied about 158 hectares (391 ac) in 2017 [3].

Strawberry yield varies depending on the production region. In 2018, the total national utilized yield was 1,295,049 mt (fresh equivalent), which generated approximately \$2,670,523,000 revenue. The average strawberry yield was 10.66 mt per hectare (29.03 t ac^{-1}) fresh equivalent, and yields ranged from 90.4 metric tons per hectare (40.3 t ac^{-1}) in California to 5.5 metric tons per hectare (2.5 t ac^{-1}) in New York [1]. The main reason for the yield differences across different states is climate conditions. California weather allows it to produce strawberries all year round, resulting in higher yields. At the same time, other states are limited to short growing seasons with harvest lasting approximately one to five months [3]. The short production season in states with colder climates is associated with lower prices received by growers. The retail price of strawberries is usually higher during the off-season and can even double in December. However, this gap has declined due to improved trade and increased year-round supply [4].

The demand for strawberries has increased drastically in the past decades. Consumption per capita increased from approximately 0.9 kg (2 lbs) per year in 1980 to 3.6 kg (7.9 lbs) per year in 2013. This increase is due to the awareness of strawberries' health benefits and increased strawberry availability all year round due to increased domestic production and increased imports [5]. In addition to the increase in per capita strawberry consumption, there is an upward trend in consumers' demand for locally grown food, presenting an opportunity for local producers to receive premium prices. Between 1994 and 2014, farmers' markets increased from 1755 to 8284 [6].

Most strawberries grown in the mid-south US are sold directly to local markets and contribute to agritourism activities. In a survey conducted by Virginia Cooperative Extension, 79% of strawberry growers stated that they sell their berries through pre-pick retail and U-pick outlets [7]. In the City of Virginia Beach, the highest producer of strawberries in Virginia, an estimated 20% of the yield is sold pre-picked at farmers' markets, and the remaining portion of the harvest is sold through U-Pick [8].

An increase in local food demand is an opportunity for strawberry producers to adopt technology such as high tunnels that help expand their production and extend their season. High tunnels are unheated, polyethylene-cover, greenhouse-like structures [9]. They offer protection from unfavorable weather conditions such as high wind, frost, hail, and precipitation, and they are reported to extend the harvest season for many crops, allowing the growers to gain early entry into the market when consumers are most excited about berry consumption and picking [9]. A study in Utah reported that high tunnels advanced June-bearing strawberry production by 4 to 5 weeks [10]. The noted benefits from high tunnels, coupled with their low installation costs compared to other protected structures, have stimulated interest among high-value crop growers [11].

Previous analyses of high tunnel vs. open-field production have yielded mixed results. Some studies in North America reported the role of high tunnels in improving the productivity and quality of high-value crops and the possibility of allowing producers to access offseason premium prices [12–15]. However, a study in Tennessee [16] reported lower net revenue from strawberries produced under a high tunnel compared to open- field production. An Arkansas study of primocane-fruiting blackberries found that open- field production was more profitable than high tunnel production [17]. A Michigan study of high tunnel and open-field raspberry production found higher insect pest pressure under high tunnels compared to open-field [18]. Another study [19] notes that while increased strawberry yield would not justify the cost of a high tunnel, growers point out that the main advantage of a high tunnel is the reliability of production regardless of rain events during harvest. Further research is needed to determine how strawberry net revenues in the southeastern U.S. are affected by high tunnel production and what factors contribute to the change in net revenues. As strawberry producers explore alternatives to conventional open-field production to expand their growing season, little is known as to whether high tunnel benefits observed in other regions can be translated to the southeastern U.S. and whether potentially higher yields and price premiums obtained during the offseason would increase the revenue enough to cover initial installation costs. The effect of different strawberry cultivars on high tunnel versus open-field comparison is also unknown. We conducted this study with the primary objective of learning whether there are additional returns from strawberry cultivars grown in high tunnels compared to those in open-fields in the Southeast and specifically Virginia and what factors contribute to the change in net returns from high tunnel production. Secondly, we were interested in learning and identifying the interaction of strawberry cultivars with the high tunnel versus open-field economic returns. Finally, we were interested in learning the sensitivity of additional returns to the market price.

2. Materials and Methods

2.1. Horticultural Design

The horticultural study was conducted at Virginia Tech's Hampton Roads Agricultural Research and Extension Center (AREC) in Virginia Beach, Virginia, during the 2019/20 season. The soil was tetotum loam with a non-amended pH of 5.9 and 5.1 for the land with and without strawberry history, respectively. Lime was applied at a rate of 2241 kg ha^{-1} (2000 lb ac^{-1}) to land with strawberry history and 5043 kg ha^{-1} (4500 lb ac^{-1}) to land without strawberry history to adjust the pH of both land types to a 6.2 level. Eight strawberry cultivars were evaluated in a randomized complete block. Five cultivars were short-day (June bearing), and three were day-neutral (spring and fall-bearing). The June-bearing cultivars were 'Rocco' [20], 'Ruby June' [21], 'Camino Real' and 'Merced' [22], and 'Chandler' [23]. The day-neutral cultivars were 'San Andreas' [22], 'Sweet Ann' (Lassen Canyon Nursery; 2009), and 'Albion' [22].

The strawberry plugs of all cultivars were ordered from the same nursery. They were planted in the first week of October 2019 and transplanted in annual hill plasticulture beds with 36 cm (14 in) in-row spacing in a staggered manner. A total of 8 rows 21 m (70 ft) long and 1.8 m (6 ft) center were planted. Four rows were inside a 9 × 45 m (30 × 148 ft) high tunnel, and four others were outside in the open-field. Each cultivar was planted on a 2 m^2 (21 ft^2) block and replicated four times in both high tunnel and open-field. The total area used for the trials in each environment was 195.1 m^2 (2100 ft^2). The planted area was 78 m^2 (840 ft^2), another 78 m^2 was for the alley-ways, and the remaining 49 m^2 (420 ft^2) was the buffer area used to separate cultivars in the trial. We ignored this buffer area in our analysis because it would not be present on a commercial operation. Therefore, we considered the area used for the trials in each environment to be 156 m^2 (1680 ft^2).

Temperature and plant health were monitored throughout the study. Temperature probes were placed at canopy levels and 15.2 cm (6 in) depth under the soil (root zone) in each bed row. Plant runner counts were recorded monthly for each replicate using a rating from 1 to 5, 1 being for rare runners and 5 for the most runners. Similarly, plant health rating was done on a monthly basis using a rating from 0 to 10, 0 meaning all plants dead in a replicate, and 10 meaning extremely vigorous and healthy appearing plants in a replicate. Strawberry plant development was monitored by measuring plant canopy diameter early in the growing season, mid-season, and toward the end of the growing season. Field plots were harvested two to three times per week by project personnel.

2.2. Economic Analysis

Marketable and nonmarketable yields per block were recorded by harvest date for each cultivar in both high tunnel and open-field environments. We added monthly marketable yield from each of the four blocks of the same cultivar in each environment to obtain the monthly yield per cultivar. Then, we extrapolated the yield to a per hectare basis.

For each cultivar, we calculated total costs, gross revenues, and net returns (revenue generated minus production cost) per hectare for three marketing strategies in both high tunnel and open-field. In the first strategy, producers pre-pick their berries and sell them at a wholesale market (pre-pick-wholesale). In the second, they pre-pick and retail berries (pre-pick-retail); in the third strategy, consumers pick the berries for themselves (U-pick).

2.2.1. Gross Revenue

To calculate the seasonal gross revenue, we added the products of monthly marketable yield per cultivar in kilograms and the estimated monthly price per kilogram during the 2019–2020 harvest season. Estimated monthly prices differ by marketing strategy. The same monthly price is used for all cultivars with high tunnel or open-field production, except that January, February, and March monthly prices are not used for open-field production as there are no yields in those months. To account for inflation, all prices were expressed in terms of the purchasing power of money in 2020 (2020 dollars) using U.S. GDP implicit price deflators obtained from the Federal Reserve Bank of St Louis website (https://fred.stlouisfed.org, accessed on 3 March 2021). To convert dollars for a cost or revenue item from a given quarter and year to 2020 dollars, we find the ratio of the deflator for the second quarter of 2020 and the deflator of the quarter in the year under consideration. This ratio is multiplied by a price reported in that year to convert it to 2020 dollars. The result was a set of prices expressed in 2020 dollars.

To estimate the monthly price received by producers in the southeastern US at the wholesale market, we found the 4-year average North Carolina strawberry prices (expressed in 2020 dollars) at the Baltimore Terminal Market (2017–2020). The terminal market price dataset was obtained from the USDA (AMS) website (https://www.marketnews.usda.gov, accessed on 3 March 2021). This dataset reported weekly low, high, mostly low, and mostly high prices per flat 8 0.45 kg (1 lb) container with a lid. We took the average of the mostly high and mostly low prices for each month and converted these average prices to 2020 dollars, as described above. We took the average of the mostly high and mostly low prices in order to capture the range of prices observed without being overly influenced by outliers that might have been introduced with inclusion of the high and low price categories.

The USDA dataset for Baltimore Terminal Market price contained only information on April through June prices. We obtained a dataset of monthly southeastern US retail prices per kilogram for five years (2016–2020) from the USDA website (https://www. marketnews.usda.gov, accessed on 4 March 2021). We converted all prices to 2020 dollars as described above and then used the average price for each month over the five years to estimate wholesale prices for January through March. First, we calculated the average percent retail margin over the terminal market price for April through June. The resulting average percent retail margin was used to estimate the average wholesale terminal market price for January, February, and March based on the retail prices for each month.

Prices for pre-pick retail and U-pick operations were obtained by finding an average of prices reported by local Virginia producers who responded to an online survey administered in 2021. Respondents market their berries via pre-pick retail or U-pick outlets. Survey respondents were asked to report their prices as dollars per pound or dollars per 4-quart basket.

2.2.2. Cost

We used the costs recorded in trials at the AREC in Virginia Beach. Since the trials were conducted on a 156 m^2 (1680 ft^2) area with eight cultivars under high tunnel and open-field, we extrapolated the costs to a per-hectare basis. Some production labor costs were based on North Carolina State extension budgets, rather than experimental observations, to more accurately reflect production conditions on a commercial operation [24]. In addition, other supplementary cost information was also obtained from the North Carolina extension budgets [24]. All costs were expressed in 2020 dollars using the procedure described above for prices.

Costs are divided into production and overhead. Production costs were reported by strawberry production stages: land preparation, transplant, dormant, harvest, and post-harvest. Except for the harvest stage, all production stages started and ended around the same time in the open-field and the high tunnel. Costs for all activities that occurred between the first and the last harvest days were classified in the harvest production stage except the labor cost spent for the actual picking of the berries, which was classified in a separate harvest labor category.

For each production stage, reported production costs include labor, materials, and machinery variable costs. Production costs of the high tunnel differ from those of the open-field. Cultivars grown within the open-field (high tunnel) have the same costs with the exception of harvest labor costs, which vary by yield and by the marketing strategy. We used harvest labor reported in the trials for the pre-pick wholesale and retail operations, and we assumed the U-pick harvest labor to be 20% of the pre-pick harvest labor.

Overhead costs include interest, depreciation, and repairs for all irrigation, machinery, and high tunnel infrastructure. Overhead costs differ between the high tunnel and open-field but are the same for all cultivars within the open-field (high tunnel). The high tunnel structure cost was depreciated for ten years based on the Internal Revenue Services' 2012 Farmers Tax Guide, which states that horticultural structures' lifetime is ten years [25]. Maintenance and interest costs were, respectively, four and five percent of the average costs of the structure.

2.2.3. Net Revenues

Net revenues per hectare from eight strawberry cultivars under each marketing strategy and under open-field and high tunnel production were estimated by subtracting the total per hectare cost from gross revenues.

3. Results

3.1. Gross Revenues

3.1.1. Budget Analysis

Gross revenues depended on cultivars, production environment, and marketing strategy. Figure 1 shows monthly revenue from each cultivar in the high tunnel (a) and open-field (b) for a pre-pick-wholesale price. 'Camino Real' generated the highest revenue in the high tunnel with annual revenue of \$73,287 ha^{-1} (\$29,659 ac^{-1}), followed by 'Merced' and 'Ruby June' with gross revenues of \$58,839 ha^{-1} (\$23,812 ac^{-1}) and \$56,366 ha^{-1} (\$22,811 ac^{-1}), respectively. 'Albion' generated the lowest revenue of \$36,578 ha^{-1} (\$14,803 ac^{-1}). In the open-field, 'Rocco' generated the highest revenue of \$106,243 ha^{-1} (\$42,996 ac^{-1}) followed by 'Sweet Ann' and 'Chandler' with revenues of \$94,715 ha^{-1} (\$38,331 ac^{-1}) and \$89,925 ha^{-1} (\$36,392 ac^{-1}), respectively. Similar to the high tunnel result, 'Albion' generated the lowest revenue of \$47,519 ha^{-1} (\$19,231 ac^{-1}).

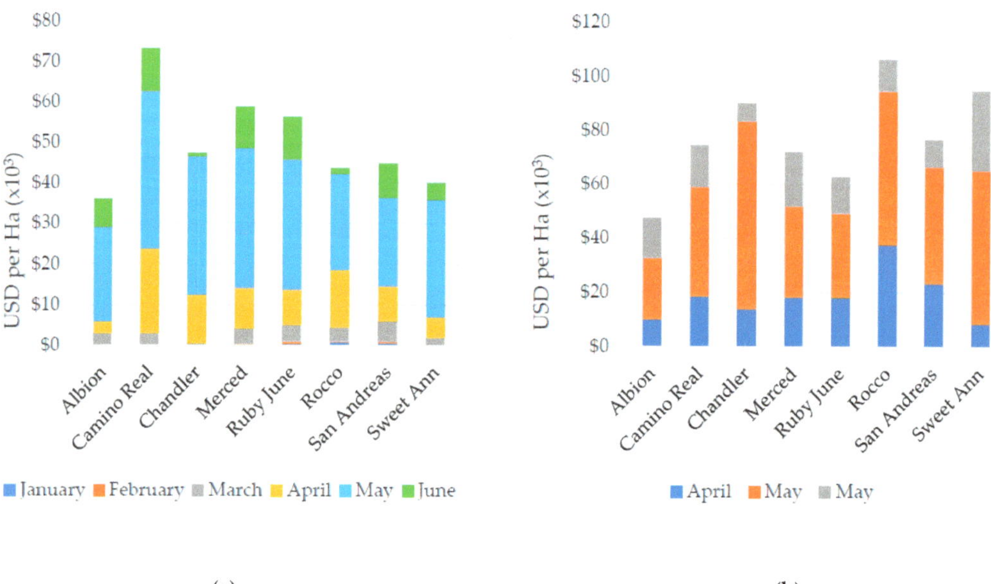

Figure 1. High tunnel (**a**) and open-field (**b**) gross revenues in USD per hectare for eight strawberry cultivars marketed at a wholesale price.

Overall, revenues from cultivars grown in the open-field were higher than those from high tunnel regardless of whether producers marketed at a wholesale, retail, or U-pick price (Table A1). Depending on the individual cultivar, per hectare revenues from the open-field were between \$1196 and \$62,440 ha^{-1} (\$484 and \$25,269 ac^{-1}) higher than those from the high tunnel for the pre-pick wholesale marketing strategy. The wholesale price, which was the 5-year average price in southeastern U.S stores, varied between \$4.59 and \$5.67 kg^{-1} (\$2.08 and \$2.57 lb^{-1}), depending on the month. The revenue difference became even wider when strawberries were marketed at a retail (\$9.48 kg^{-1} or \$4.30 lb^{-1}) or U-pick (\$6.97 kg^{-1} or \$3.16 lb^{-1}) price. The differences ranged between \$2430 and \$118,803 ha^{-1} (\$983 and \$48,078 ac^{-1}) for the pre-pick retail marketing strategy and from \$1786–\$87,306 ha^{-1} (\$723 to \$35,332 ac^{-1}) for U-pick. On average, the revenue was \$27,873 ha^{-1} (\$11,280 ac^{-1}), \$53,233 ha^{-1} (\$21,543 ac^{-1}), and \$39,120 ha^{-1} (\$15,831 ac^{-1}) higher in open-field than in the high tunnel for pre-pick-wholesale, pre-pick-retail, and U-pick marketing strategies, respectively.

3.1.2. Yield Analysis

The high tunnel allowed harvest before the normal harvest season (before April), as harvest for some cultivars started as early as January. However, the yields were generally low compared to the open-field. The total yields for different cultivars ranged between 9676–21,379 kg ha^{-1}) (8634 and 19,078 lb ac^{-1}) in the open-field and from 7550–14,865 kg ha^{-1} (6737 to 13,265 lb ac^{-1}) in the high tunnel. The average yield per hectare was 1.4 times higher in the open-field compared to the high tunnel. The yields were slightly skewed to the right in the high tunnel (Figure 2a) and normally distributed in the open-field (Figure 2b).

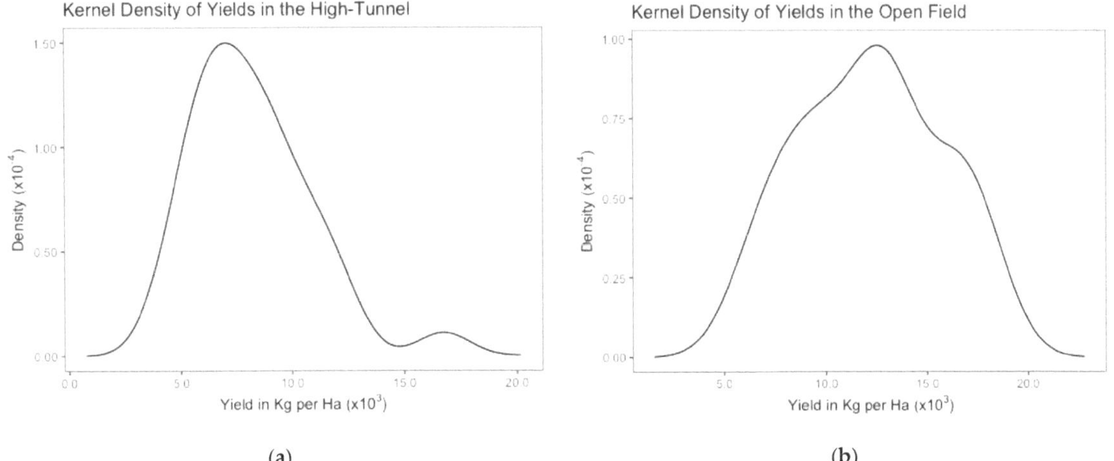

Figure 2. Kernel density estimates for high tunnel (**a**) and open-field yields (**b**).

A simple linear regression analysis sheds additional light on the effects of cultivar and production environment, namely open-field and high tunnel, on yield. Tables 1 and 2 report the estimates from ordinary least square regressions (OLS) with yields as the dependent variable and cultivar as the independent variable. Using ordinary least squares is justified because data were obtained from experiments where other factors, such as weather and soil, were controlled, and endogeneity should not be an issue. Ordinary least squares estimation method assumes that errors have a constant variance, homoskedasicity. To ensure this assumption was met in regression models in Tables 1 and 2, we conducted the Breusch-pagan heteroskedasticity test. We failed to reject the null hypothesis that homoskedasticity is present in both high tunnel ($p = 0.21$) and open-field ($p = 0.43$) models.

The intercepts in Tables 1 and 2 indicate an estimate of the average yields for the base variety, 'Chandler' under high tunnel and open-field conditions, respectively. Estimates for other varieties indicate how much the yield of each variety varies from the estimate for 'Chandler'. In the high tunnel, the average yield for 'Chandler' was 7725 kg ha^{-1} (6891 lb ac^{-1}), as indicated by the intercept estimate in Table 1. Only 'Camino Real' had a statistically significant higher yield than 'Chandler' ($p = 0.013$). On average, its yield was 4167 kg ha^{-1} (3717 lb ac^{-1}) more than 'Chandler's'. Similarly, the yield for 'Chandler' in the open-field was 14,527 kg ha^{-1} (12,959 lb ac^{-1}), as indicated by the estimate for open-field in Table 2, almost twice the yield reported in the high tunnel. The open-field average yields for 'Camino Real' ($p < 0.1$), 'Ruby June' ($p < 0.01$), 'Albion' ($p < 0.001$), and 'Merced' ($p < 0.1$) were lower than those for 'Chandler'.

Table 1. Estimated average yield (kg ha^{-1}) from strawberry cultivars produced in the high tunnel with 'Chandler' as the base category.

	Estimate	Std. Error	t Value	Pr (> | t |)
Intercept	7725	1098	7.04	2.82×10^{-7} ***
Albion	−1686	1552	−1.09	0.288
Camino Real	4167	1552	2.68	0.013 *
Merced	1917	1552	1.24	0.229
Ruby June	1520	1552	0.98	0.337
Rocco	−645	1552	−0.42	0.681
San Andreas	−371	1552	−0.24	0.813
Sweet Ann	−1149	1552	−0.74	0.466

Signif. codes: '***' 0.001, '*' 0.05.

Table 2. Estimated average yield (kg ha^{-1}) from strawberry cultivars produced in the open-field with 'Chandler' cultivar as the base category.

	Estimate	Std. Error	t Value	Pr (> | t |)
Intercept	14,527	1215	11.95	1.36×10^{-11} ***
Albion	−6807	1719	−3.96	0.000582 ***
Camino Real	−2963	1719	−1.72	0.097579
Merced	−3252	1719	−1.89	0.070599
Ruby June	−5118	1719	−2.98	0.006545 **
Rocco	1351	1719	0.79	0.439687
San Andreas	−2576	1719	−1.50	0.146967
Sweet Ann	919	1719	0.54	0.597786

Signif. codes: '***' 0.001, '**' 0.01.

3.2. Cost

3.2.1. Production Cost

Overall, the production cost, excluding harvest labor for picking berries, was higher in the high tunnel than in the open-field across all strawberry production stages except for the pre-harvest stage, as presented in Figure 3a,b. Transplanting was the costliest production stage in both high tunnel and open-field, and the post-harvest stage was the least costly. The transplanting production stage costs $12,414 and $11,458 ha^{-1} ($5024 and $4637 ac^{-1}) in the high tunnel and open-field, respectively. The least costly stage was post-harvest. It costs $115 ha^{-1} ($46.50 ac^{-1}) in both high tunnel and open-field. The total per hectare production cost (excluding harvest labor for picking berries) was $30,327 ($12,273 ac^{-1}) in the high tunnel and $25,829 ha^{-1} ($10,453 ac^{-1}) in the open-field.

Expenses for each production stage were categorized into equipment, materials, and labor. The materials category was the costliest in both high tunnel and open-field. It costs $16,002 and $14,228 ha^{-1} ($6476 and $5758 ac^{-1}) in the high tunnel and open-field, respectively, equivalent to 52.8% and 55.1% of the production costs in the high tunnel and open-field, respectively. Equipment cost was the least expensive category, $2270 ha^{-1} ($919 ac^{-1}) and $2241 ha^{-1} ($907 ac^{-1}) in the high tunnel and open-field, respectively.

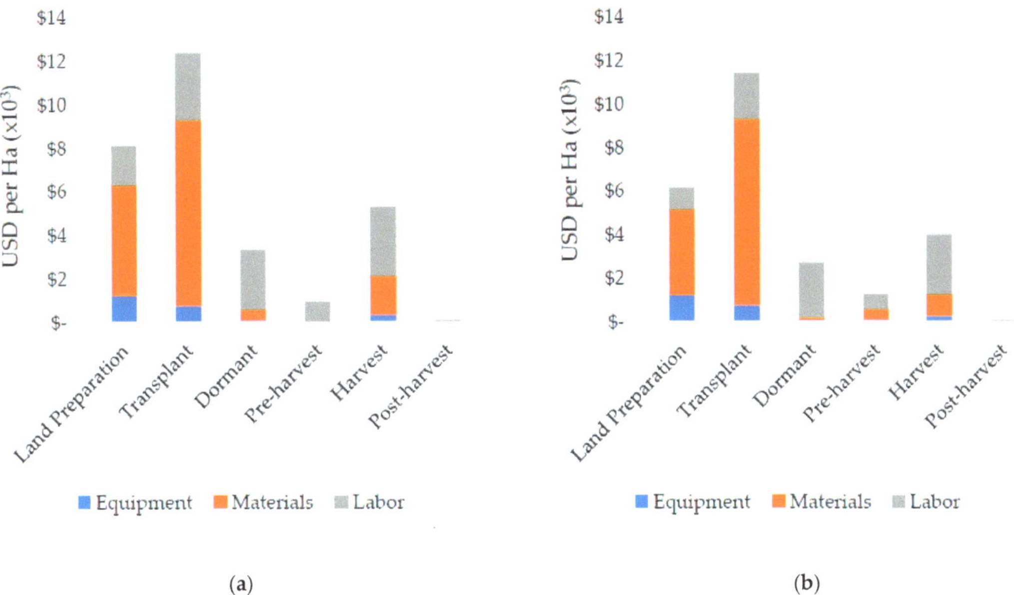

(a)   (b)

Figure 3. Estimated production costs in USD per hectare for every strawberry production stage in the high tunnel (**a**) and open-field (**b**). Labor for the 'harvest production stage' does not include labor for picking berries.

3.2.2. Administrative and Cumulative Cost

The annual administrative cost, including real estate taxes, net land rent, and overhead, was $378 ha^{-1} ($153 ac^{-1}) in both high tunnel and open-field. Adding this cost to the total high tunnel and open-field production cost, we obtained a cumulative cost of $30,705 ha^{-1} ($12,426 ac^{-1}) and $26,209 ha^{-1} ($10,606 ac^{-1}), respectively, excluding harvest labor. We added $65,331 ha^{-1} ($26,439 ac^{-1}) to the high tunnel cumulative cost for annual ownership cost of the structure resulting in a cumulative cost of $96,039 ha^{-1}, ($38,865 ac^{-1}) slightly more than three and a half times the cumulative cost of producing strawberries in the open-field.

3.2.3. Harvest Labor Cost

The costs of growing different cultivars under the same production environment differ by harvest labor costs which, in turn, vary by marketing strategy chosen by producers and by cultivar. Figure 4 compares the harvest labor cost per hectare in the high tunnel and open-field for each individual cultivar and pre-pick marketing strategy. For all cultivars, harvest labor costs were higher in the high tunnel than in the open-field. The largest difference in harvest labor was recorded for 'Camino Real'. The average harvest labor in the high tunnel was a little over 1.4 times the average in open-field for a pre-pick operation. This was somewhat unexpected based on the fact that high tunnel strawberry cultivars had a lower yield compared to cultivars in the open-field. This higher harvest cost in the high tunnel can be explained by the increased number of harvesting times because harvest started earlier in the high tunnel compared to the open-field. The patterns in the harvest labor for a U-pick operation were similar to those for a pre-pick operation. The only difference is that U-pick harvest labor provided by the grower was assumed to be 20 percent of that in the open-field.

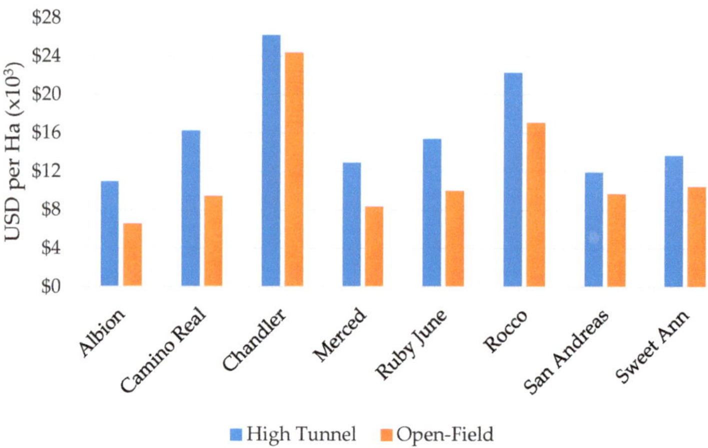

Figure 4. Harvest labor cost in USD per hectare for high tunnel and open-field cultivars and for a pre-pick operation.

The harvest labor cost differs by strawberry cultivar. Yields partially explain cost differences. Three of the four highest yielding high-tunnel cultivars had the highest harvest labor costs for the high tunnel: 'Chandler' ($26,170 ha^{-1} or $10,591 ac^{-1}), 'Camino Real' (16,215 ha^{-1} or $6562 ac^{-1}), and 'Ruby June' ($15,385 ha^{-1} or $6226 ac^{-1}). 'Albion' had the lowest high-tunnel harvest cost ($10,924 ha^{-1} or $4421 ac^{-1}) and the lowest yield. However, 'Rocco' had the second-highest high tunnel labor cost ($22,289 ha^{-1} or $9020 ac^{-1}) but the sixth-highest yield.

The three highest-yielding open-field cultivars had the highest open-field harvest labor costs: 'Chandler' ($24,379 ha^{-1} or $9867 ac^{-1}), 'Rocco' ($17,097 ha^{-1} or $6919 ac^{-1}), and 'Sweet Ann' (10,472 ha^{-1} or $4238 ac^{-1}). 'Albion' had the lowest open-field harvest labor cost of $6538 ha^{-1} ($2646 ac^{-1}) and the lowest yield. However, 'Ruby June' had the fourth highest labor cost ($9987 ha^{-1} or $4042 ac^{-1}) but the seventh highest open-field yield.

The cumulative cost (including harvest labor) for a pre-pick operation ranged between $106,963 and $122,209 ha^{-1} ($43,286 and $49,456 ac^{-1}) in the high tunnel, and $32,747 and $50,590 ha^{-1} ($13,252 and $20,473 ac^{-1}) in the open-field depending on strawberry cultivars. Similarly, for a U-pick operation, total cost ranged between $98,223 and $101,273 ha^{-1} ($39,750 and $40,984 ac^{-1}) in the high tunnel and $27,516–$31,085 ha^{-1} ($11,135 and $12,580 ac^{-1}) in the open-field (Table A2).

3.3. Net Revenues

Net revenues depended on the production environment, strawberry cultivars, and market price. Overall, producing strawberries in the high tunnel was not profitable (Table A3). When strawberries were marketed at wholesale price, all cultivars in the high tunnel generated negative net revenues (Figure 5a). In comparison, all cultivars in the open-field generated positive net returns (Figure 5a and Table A3). Net revenues from cultivars produced in the high tunnel ranged between −$74,638 and −$38,965 ha^{-1} (−$30,205 and −$15,768 ac^{-1}) while those from open-field cultivars ranged between $14,774 and $62,940 ha^{-1} ($5979 and $25,471 ac^{-1}). On average, open-field cultivars generated $101,893 ha^{-1} ($41,235 ac^{-1}) higher net revenues compared to the high tunnel.

Similarly, when producers marketed their strawberries at a retail price (pre-pick retail) and when consumers picked the berries from the field (U-pick), all high tunnel cultivars generated a negative net return, except for 'Camino Real' and 'Merced'. The net revenue from 'Camino Real' was $28,689 ha^{-1} ($11,610 ac^{-1}) and $4295 ha^{-1} ($1738 ac^{-1}) at pre-pick-retail and U-pick prices, respectively, and the net revenue from 'Merced' was

5338 ha^{-1} (2160 ac^{-1}) at the pre-pick retail price. In contrast, all open-field cultivars generated a positive return (Figure 5b,c and Table A3). In both marketing strategies, the loss from producing strawberries in the high tunnel was smaller than that incurred when berries were marketed at a wholesale price (pre-pick wholesale), but the difference between high tunnel and open-field net revenue became wider. On average, the per hectare net revenue from open-field cultivars exceeded that from the high tunnel by $127,253 \text{ ha}^{-1}$ ($51,498 \text{ ac}^{-1}$) for pre-pick retail and $109,788 \text{ ha}^{-1}$ ($44,430 \text{ ac}^{-1}$) for U-pick.

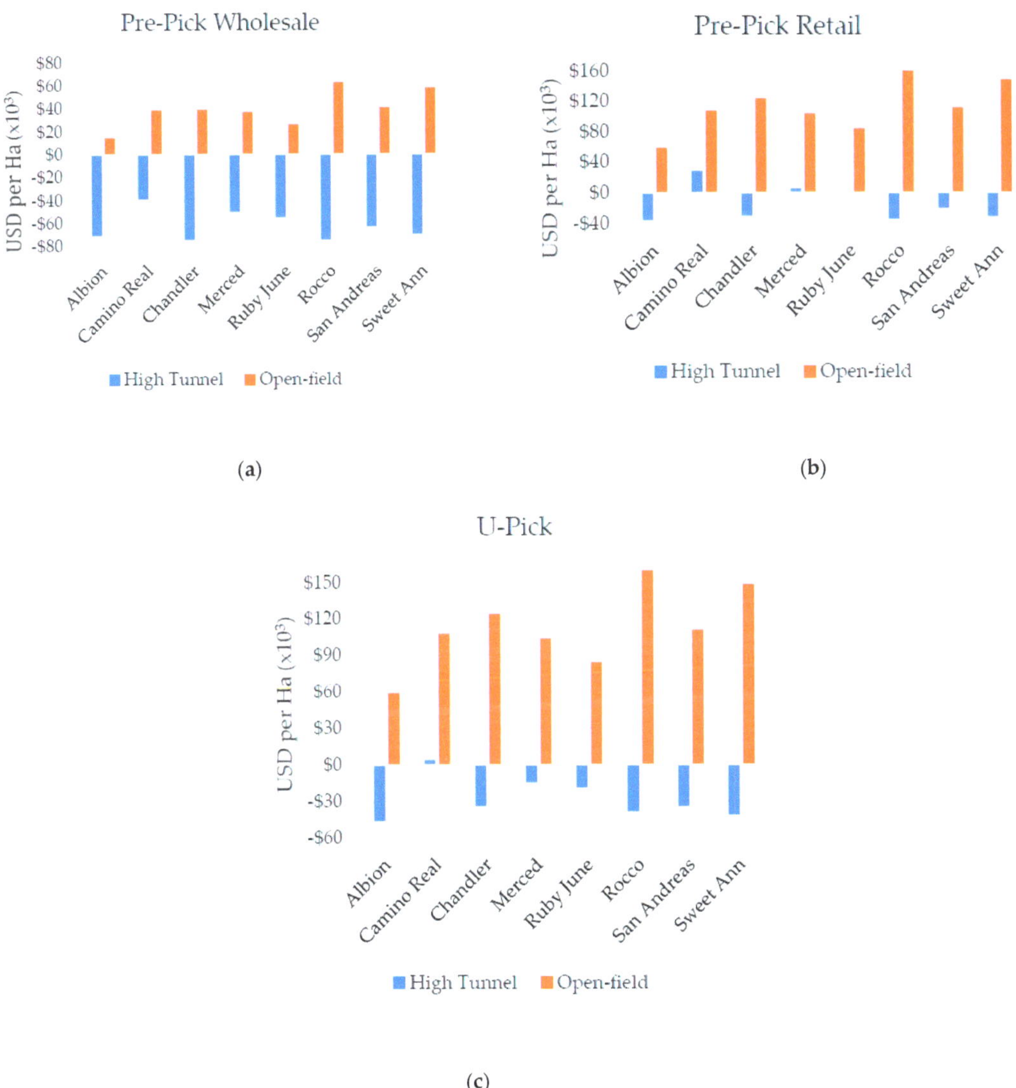

Figure 5. Net revenues in USD per hectare from strawberries produced under a high tunnel and in the open-field and marketed at a (**a**) pre-pick wholesale, (**b**) pre-pick retail, and (**c**) U-pick price.

4. Discussion

Several studies conducted in different regions in North America reported higher net revenues from producing fruits and vegetables in high tunnels compared to the open-field [26]. High tunnels extended the harvest season and increased marketable yields resulting in higher net returns. In contrast, our study finds lower net returns from high tunnel production for all marketing strategies. Lower net returns in the high tunnel compared to the open-field can be attributed to two main reasons. First, the high tunnel production cost was higher than the cost of the open-field cultivars, as discussed in the previous section (Table A2). A major reason for the higher cost is the overhead cost of the high tunnel structure, which makes up almost half of the total cost of high tunnel production. Second, as a result, the early season strawberry yields and revenues generated from high tunnel production compared to open-field production were not enough to offset the higher cost of owning the high tunnel structure for any of the three marketing prices used in this study. In fact, average yields from the high tunnel were lower than from the open-field (Tables 1 and 2).

The lower yields in the high tunnel compared to the open-field corroborates the findings from a Tennessee study that spring-bearing open-field strawberries produced higher yields than winter and spring-bearing high-tunnel strawberries [16]. One possible reason for lower high tunnel yields during our trials is herbicide and disease damage resulting from cooler conditions in winter and warmer conditions in spring. Early fruiting in the high tunnel when conditions were cooler favored the development of diseases in the high tunnel at the time when there were no fruits in the open-field. Forcing strawberries to bear fruit during the winter months can be a drawback as yield is reduced compared to traditional spring production [27].

The warm conditions in the high tunnel during the spring compared to the open-field environment favored pests and led to lower yields. This is similar to what Martin reported that warm winter temperatures provided insects an overwintering location in the high tunnel and caused more damage to the berries during winter and spring production [16]. Leach and Isaacs [18] also reported higher pest pressure in high tunnel raspberries compared to open-field production while Ingwell et al. [28] also observed a higher pest prevalence in high tunnels compared to open-field plots in a study of tomatoes, cucumbers, and broccoli. Rodents are also attracted to warm conditions in the high tunnel during winter and feed on high tunnel crops [29]. Overhead irrigation was used in our study to protect the plants from frost, insect damage, and diseases, but it is clear that the yield loss was enlarged compared to open-field production.

Galitano and Miles [11] also reported a lower profit from lettuce produced in the high tunnel than in the open-field. The net revenue from lettuce produced in the high tunnel was 30% less per square meter than the net revenue from producing in the open-field. The higher yield from the high tunnel observed in their study was insufficient to offset the increased costs. This finding suggests that even higher yields in the high tunnel would not have been a sufficient condition to guarantee higher net revenues at a low wholesale price, but it might have at higher prices for other marketing strategies. Results from previous studies as well as this study suggest greater attention needs to be given to managing strawberry production under high tunnels especially with respect to pest management.

Focusing on the open-field, where net-revenues were always positive, 'Rocco' and 'Sweet Ann' would be the best choices for U-pick and pre-pick retail due to higher net revenues (Figure 5b,c). Their yields are higher than 'Chandler' yields although the differences are not statistically significant (Table 1). However, these cultivars are not suitable for shipping making them less suited to pre-pick wholesale operations. Additionally, although data from research and on-farm trials shows that yield from 'Rocco' is similar to that of 'Chandler', it is still a new cultivar [20]. 'Sweet Ann' does not tolerate rain events very well due to its thin fruit skin. It did well in the outdoor environment in our trial because low rain events and intensity were recorded this season. During harvest seasons when there are high rainfall events and intensity, this cultivar may not yield the same amount of marketable yield. 'Chandler', which ranked third in open-field net revenues for pre-pick retail and U-pick, has been grown by southeastern growers for a long time, and it has been among the top cultivars with high yields [30] making it favorable for U-pick and pre-pick retail operations. 'Camino Real' ranked fifth in open-field net revenues. Previous studies reported that it can yield as well or better than 'Chandler' in different regions of Eastern Virginia [31]. The high yield potential, rain resistance, ease of picking, and good shipping ability makes 'Camino Real' a favorable cultivar for pre-pick wholesale.

5. Conclusions

In this paper, we study and compare cost, revenues, and net revenues from eight strawberry cultivars produced under a high tunnel and in the open-field in the Commonwealth of Virginia. We find that producing in the high tunnel was not profitable for three marketing strategies: pre-pick with wholesale marketing, pre-pick with retail marketing, and U-pick whereby consumers pick strawberries.

The total production cost (including harvest, ownership, and administrative costs) for the high tunnel averaged \$112,223 (\$45,415 ac^{-1}) and \$99,276 (\$40,175 ac^{-1}) for pre-pick and U-pick, respectively, and for the open-field averaged \$38,204 (\$15,460 ac^{-1}) and \$28,608 (\$11,577 ac^{-1}) for pre-pick and U-pick, respectively. Revenues from all eight cultivars were lower in the high tunnel than in the open-field, regardless of whether strawberries were sold at a wholesale, retail, or U-pick price. This is a result of lower marketable yields from the high tunnel compared to the open-field. High tunnel net revenues were negative for all cultivars and marketing strategies, except for 'Camino Real' and 'Merced'. The net revenue from 'Camino Real' was positive at pre-pick-retail and U-pick prices, and the net revenue from 'Merced' was positive at a pre-pick retail price. In contrast, net revenues from open-field were positive for all cultivars and marketing strategies. On average, net revenue was \$101,893, \$127,253, and \$109,788 ha^{-1} (\$41,235, \$51,498, and \$44,430 ac^{-1}) higher in the open-field than in the high tunnel for pre-pick wholesale, pre-pick retail, and U-pick marketing, respectively.

Study results imply that growers should focus on open-field production of strawberries rather than under a high tunnel. The most promising cultivars in terms of net returns from open-field production are 'Rocco', 'Sweet Ann', 'Chandler', and 'Camino Real'. However, 'Rocco' has limitations for shipping and 'Sweet Ann' has limitations due to sensitivity to damage from rain events. A limitation to this study was that the economic data was run on a field trial based on a single growing season, i.e., 2019–2020 growing season. Replicating the study might provide additional insights into the effects of growing season conditions on the differences in yields from high tunnel and open-field production.

Further research is also warranted to develop an improved production system for strawberries under high tunnels. Disease and insect pest management methods need to be adapted to the high tunnel production environment in order to produce the increased yields needed to offset the overhead costs of the high tunnel structure. Harvest costs reported in this study may be higher than those realized by commercial growers because of experimental methods used to record our harvest labor. Labor inputs were reported for small plots, and experiment workers harvested all berries, including non-marketable yields. Harvest workers on a commercial operation may be more efficient, resulting in lower harvest labor costs, which could be corroborated with further research. Additionally, further research is needed to evaluate the relative profitability of the three marketing strategies: U-pick, retail, and wholesale. Such study should consider all marketing costs for each marketing strategy.

Author Contributions: Co-authors were involved in the manuscript preparation as follows: Conceptualization, J.B.S. and D.J.B.; Methodology, J.C.M. and D.J.B.; Software, J.C.M.; Validation, J.C.M., D.J.B. and J.B.S.; Formal Analysis, J.C.M. and D.J.B.; Investigation, J.C.M., D.J.B. and J.B.S.; Resources, D.J.B. and J.B.S.; Data Curation, J.C.M., D.J.B. and J.B.S.; Writing—Original Draft Preparation, J.C.M.; Writing—Review and Editing, J.C.M., D.J.B. and J.B.S.; Visualization, J.C.M.; Supervision, D.J.B. and J.B.S.; Project Administration, J.B.S. and D.J.B.; Funding Acquisition, J.B.S. and D.J.B. All authors have read and agreed to the published version of the manuscript.

Funding: This research was funded in part by the United States Department of Agriculture (USDA), Specialty Crop Grant Block (Grant# 418963). Additional field research support was provided by USDA NIFA Hatch Project VA-160077. Staff and faculty salary and support were provided by the Virginia Agricultural Experiment Station (Blacksburg) and USDA NIFA Hatch Project 1016661.

Institutional Review Board Statement: Not applicable.

Informed Consent Statement: Not applicable.

Data Availability Statement: The data presented in this study are available in https://ext.vt.edu/small-fruit.html (accessed on 17 November 2022).

Acknowledgments: The authors would like to thank Hampton Roads AREC farm: Robert Holtz for field assistance and Danyang Liu and Sophia Gonzales for their contribution in gathering and recording data from strawberry cultivar trials. The authors would also like to thank Barclay Poling, Roy D. Flanagan, and Stephanie Romelczyk for their feedback on the budgets used in this article, and Shamar Stewart and Ford Ramsey for their helpful suggestions on data visualization and the statistical estimations, respectively. The authors are grateful for funding received from the Virginia Tech Open Access Subvention Fund to cover article processing costs.

Conflicts of Interest: The authors declare no conflict of interest. The funders had no role in the design of the study; in the collection, analyses, or interpretation of data; in the writing of the manuscript, or in the decision to publish the results.

Appendix A

Table A1. Strawberry gross revenues (estimated yield * estimated price) in USD per hectare by cultivar and marketing strategy in the high tunnel and open-field environment.

Cultivar	Pre-Pick Wholesale			Pre-Pick Retail			U-Pick		
	High Tunnel	Open-Field	Difference	High Tunnel	Open-Field	Difference	High Tunnel	Open-Field	Difference
Albion	36,248	47,521	11,273	71,027	91,743	20,715	52,197	67,420	15,223
Camino Real	73,289	74,484	1195	140,943	143,373	2430	103,577	105,363	1786
Chandler	47,571	89,927	42,357	91,560	174,428	82,867	67,286	128,184	60,898
Merced	58,840	71,914	13,074	114,287	138,367	24,081	83,987	101,684	17,697
Ruby June	56,368	62,731	6363	109,572	120,355	10,783	80,523	88,447	7924
Rocco	43,803	106,245	62,442	83,916	202,718	118,803	61,668	148,975	87,306
San Andreas	44,927	76,613	31,686	87,158	146,793	59,635	64,051	107,876	43,825
Sweet Ann	40,122	94,719	54,596	78,117	184,670	106,553	57,407	135,711	78,304
Average	50,146	78,019	27,873	97,073	150,306	53,233	71,337	110,457	39,120
Min	36,248	47,521	1195	71,027	91,743	2430	52,197	67,420	1786
Max	73,289	106,245	62,442	140,943	202,718	118,803	103,577	148,975	87,306

Table A2. Strawberry production costs in USD per hectare (including administrative cost) by cultivar and marketing strategy in the high tunnel and open-field environment.

Cultivar	High Tunnel	Pre-Pick Open-field	Difference	High Tunnel	U-Pick Open-field	Difference
Albion	106,963	32,747	74,216	98,223	27,516	70,707
Camino Real	112,254	35,635	76,619	99,282	28,094	71,188
Chandler	122,209	50,590	71,618	101,273	31,085	70,188
Merced	108,949	34,573	74,376	98,621	27,882	70,739
Ruby June	111,424	36,196	75,228	99,116	28,206	70,909
Rocco	118,327	43,305	75,022	100,496	29,628	70,868
San Andreas	107,953	35,902	72,051	98,421	28,147	70,274
Sweet Ann	109,710	36,682	73,028	98,773	28,303	70,469
Average	112,223	38,204	74,020	99,276	28,608	70,668
Min	106,963	32,747	71,618	98,223	27,516	70,188
Max	122,209	50,590	76,619	101,273	31,085	71,188

Table A3. Strawberry net revenues in USD per hectare (gross revenues—production costs) by cultivar and marketing strategy in the high tunnel and open-field environment.

Cultivar	Pre-Pick Wholesale High Tunnel	Open-Field	Difference	Pre-Pick Retail High Tunnel	Open-Field	Difference	U-Pick High Tunnel	Open-Field	Difference
Albion	−70,714	14,774	85,489	−35,935	58,996	94,931	−46,026	39,904	85,930
Camino Real	−38,965	38,849	77,814	28,689	107,738	79,049	4295	77,268	72,973
Chandler	−74,638	39,337	113,975	−30,648	123,837	154,486	−33,986	97,099	131,085
Merced	−50,109	37,342	87,450	5338	103,795	98,457	−14,633	73,802	88,436
Ruby June	−55,056	26,535	81,591	−1851	84,160	86,011	−18,593	60,241	78,834
Rocco	−74,524	62,940	137,464	−34,411	159,413	193,825	−38,828	119,347	158,174
San Andreas	−63,026	40,711	103,737	−20,795	110,891	131,686	−34,370	79,728	114,099
Sweet Ann	−69,587	58,037	127,624	−31,593	147,988	179,581	−41,366	107,408	148,774
Average	−62,077	39,816	101,893	−15,151	112,102	127,253	−27,938	81,850	109,788
Min	−74,638	14,774	77,814	−35,935	58,996	79,049	−46,026	39,904	72,973
Max	−38,965	62,940	137,464	28,689	159,413	193,825	4295	119,347	158,174

References

1. National Agricultural Statistics Service, USDA. *Noncitrus Fruits and Nuts 2018 Summary 06/18/2019*; USDA: Washington, DC, USA, 2019; p. 101.
2. Samtani, J.B.; Rom, C.R.; Friedrich, H.; Fennimore, S.A.; Finn, C.E.; Petran, A.; Wallace, R.W.; Pritts, M.P.; Fernandez, G.; Chase, C.A.; et al. The Status and Future of the Strawberry Industry in the United States. *HortTechnology* **2019**, *29*, 11–24. [CrossRef]
3. Agricultural Marketing Resource Center. Strawberries. Available online: https://www.agmrc.org/commodities-products/fruits/strawberries (accessed on 31 August 2021).
4. USDA ERS—Charts of Note. Available online: https://www.ers.usda.gov/data-products/charts-of-note/charts-of-note/?page=3&topicId=14849 (accessed on 31 August 2021).
5. U.S. Strawberry Consumption Continues to Grow. Available online: http://www.ers.usda.gov/data-products/chart-gallery/gallery/chart-detail/?chartId=77884 (accessed on 31 August 2021).
6. Vogel, S. Number of U.S. Farmers' Markets Continues to Rise. Available online: http://www.ers.usda.gov/data-products/chart-gallery/gallery/chart-detail/?chartId=77600 (accessed on 14 December 2021).
7. Christman, J.; Samtani, J.B. *A Survey of Strawberry Production Practices in Virginia*, 1st ed.; Virginia Cooperative Extension: Blacksburg, VA, USA, 2019.
8. Strawberry Fact Sheet 2020. Available online: https://www.vbgov.com/government/departments/agriculture/Documents/Fact%20Sheets/Strawberry%20Fact%20Sheet%202020.pdf (accessed on 1 September 2021).

9. Kaiser, C.; Ernst, M. *High Tunnel Overview*; Cooperative Extension Serivice; College of Agriculture, Food and Environment, University of Kentucky: Lexington, KY, USA, 2021; Available online: https://www.uky.edu/ccd/sites/www.uky.edu.ccd/files/hightunneloverview.pdf (accessed on 10 September 2021).
10. Rowley, D.; Black, B.L.; Drost, D.; Feuz, D. Early-season Extension Using June-bearing 'Chandler' Strawberry in High-elevation High Tunnels. *HortScience* **2010**, *45*, 1464–1469. [CrossRef]
11. Galinato, S.P.; Miles, C.A. Economic Profitability of Growing Lettuce and Tomato in Western Washington under High Tunnel and Open-field Production Systems. *HortTechnology* **2013**, *23*, 453–461. [CrossRef]
12. Maughan, T.L.; Curtis, K.R.; Black, B.L.; Drost, D.T. Economic Evaluation of Implementing Strawberry Season Extension Production Technologies in the U.S. Intermountain West. *HortScience* **2015**, *50*, 395–401. [CrossRef]
13. Hecher, E.A.D.S.; Falk, C.L.; Enfield, J.; Guldan, S.J.; Uchanski, M.E. The Economics of Low-cost High Tunnels for Winter Vegetable Production in the Southwestern United States. *HortTechnology* **2014**, *24*, 7–15. [CrossRef]
14. Waterer, D. Yields and Economics of High Tunnels for Production of Warm-season Vegetable Crops. *HortTechnology* **2003**, *13*, 339–343. [CrossRef]
15. Lamont, W.J. Overview of the Use of High Tunnels Worldwide. *HortTechnology* **2009**, *19*, 25–29. [CrossRef]
16. Martin, J. The Influence of Organically Managed High Tunnel and Open Field Production Systems on Strawberry (Fragaria x ananassa) Quality and Yield, Tomato (Solanum lycopersicum) Yield, and Evaluation of Plastic Mulch Alternatives. Master's Thesis, University of Tennessee, Knoxville, TN, USA, May 2013. Available online: https://trace.tennessee.edu/utk_gradthes/1642 (accessed on 2 September 2021).
17. Rodríguez, H.G.; Popp, J.; Thomsen, M.; Friedrich, H.; Rom, C.R. Economic Analysis of Investing in Open-field or High Tunnel Primocane-fruiting Blackberry Production in Northwestern Arkansas. *HortTechnology* **2012**, *22*, 245–251. [CrossRef]
18. Leach, H.; Isaacs, R. Seasonal Occurrence of Key Arthropod Pests and Beneficial Insects in Michigan High Tunnel and Field Grown Raspberries. *Environ. Entomol.* **2018**, *47*, 567–574. [CrossRef] [PubMed]
19. Demchak, K. Small Fruit Production in High Tunnels. *HortTechnology* **2009**, *19*, 44–49. [CrossRef]
20. Fernandez, G.; Pattison, J.; Perkins-Veazie, P.; Ballington, J.R.; Clevinger, E.; Schiavone, R.; Gu, S.; Samtani, J.; Vinson, E.; McWhirt, A.; et al. 'Liz' and 'Rocco' Strawberries. *HortScience* **2020**, *55*, 597–600. [CrossRef]
21. Lassen Canyon Nursery. Ruby June Studies by Barclay Poling. 9 June 2017. Available online: https://www.lassencanyonnursery.com/ruby-june-studies-barclay-poling/ (accessed on 2 September 2021).
22. Office of Research. The UC Patented Strawberry Cultivars. Available online: https://research.ucdavis.edu/industry/ia/industry/strawberry/cultivars/ (accessed on 2 September 2021).
23. Anonymous. Strawberry Varieties Released. Plantbreeding, 10 May 2017. Available online: https://plantbreeding.ucdavis.edu/strawberry-varieties-released (accessed on 2 September 2021).
24. Rysin, O.; Poling, B.; Louws, F. Budgets/Cost Estimates. Available online: https://strawberries.ces.ncsu.edu/strawberries-budgets/ (accessed on 2 September 2021).
25. Internal Revenue Service. Farmer's Tax Guide for Use in Preparing 2012 Returns. 23 October 2012. Available online: https://www.irs.gov/pub/irs-prior/p225--2012.pdf (accessed on 14 December 2021).
26. Kadir, S.; Carey, E.; Ennahli, S. Influence of High Tunnel and Field Conditions on Strawberry Growth and Development. *HortScience* **2006**, *41*, 329–335. [CrossRef]
27. Dufault, R.J.; Ward, B.K. Further Attempts to Enhance Forced 'Sweet Charlie' Strawberry Yield through Manipulation of Light Quality in High Tunnels. *Int. J. Fruit Sci.* **2009**, *9*, 409–418. [CrossRef]
28. Ingwell, L.L.; Thompson, S.L.; Kaplan, I.; Foster, R.E. High tunnels: Protection for rather than from insect pests? *Pest Manag. Sci.* **2017**, *73*, 2439–2446. [CrossRef] [PubMed]
29. Maughan, T.; Ernst, T.; Black, B.; Drost, D. *Defending the Castle: Integrated Pest Management in High Tunnel Strawberries*; Utah State University Extension: Logan, UT, USA, 2012; Available online: https://digitalcommons.usu.edu/extension_curall/1054 (accessed on 2 September 2021).
30. Flanagan, R.D.I. Strawberry Cultivar Evaluation on Farms Utilizing Conventional Growing Methods in the Coastal Plain of Virginia. May 2017. Available online: https://vtechworks.lib.vt.edu/handle/10919/78317 (accessed on 2 September 2021).
31. Flanagan, R.D.; Samtani, J.B.; Manchester, M.A.; Romelczyk, S.; Johnson, C.S.; Lawrence, W.; Pattison, J. On-farm Evaluation of Strawberry Cultivars in Coastal Virginia. *HortTechnology* **2020**, *30*, 789–796. [CrossRef]

Article

Vermicompost and Vermicompost Leachate Application in Strawberry Production: Impact on Yield and Fruit Quality

Ranko Čabilovski, Maja S. Manojlović, Boris M. Popović, Milivoj T. Radojčin, Nenad Magazin, Klara Petković *, Dragan Kovačević and Milena D. Lakićević

Faculty of Agriculture, University of Novi Sad, Trg Dositeja Obradovića 8, 21 000 Novi Sad, Serbia
* Correspondence: klara.petkovic@polj.uns.ac.rs

Abstract: Recycling organic waste is most important for preserving natural resources. The research objective was to quantify the effect of the application of vermicompost and vermicompost leachate on the yield and quality of strawberries and compare it with a standard fertilization program with mineral fertilizers during a 3-year production cycle. Five fertilization treatments were studied: control—without fertilizer (Ø); vermicompost (V); vermicompost + foliar application of vermicompost leachate (VL); vermicompost leachate through fertigation and foliar application (L); and mineral NPK fertilizers (NPK). The application of V positively affected strawberry yield only in the first year. In all three years of fruiting, the highest yield was measured for NPK treatment. In the first year, fertilization had no effect on fruit quality, while in the second and third years, the application of leachate led to a significantly higher concentration of total soluble solids, total anthocyanins, antioxidant activity of the fruit, and a lower concentration of total acid. Strawberries are grown for a two- or three-year production cycle, so the application of V and VL cannot maintain the yield level as was with the application of mineral NPK fertilizers. The quality of strawberry fruit, however, can be improved significantly.

Keywords: fertilizers; mineral composition; antioxidant activity

Citation: Čabilovski, R.; Manojlović, M.S.; Popović, B.M.; Radojčin, M.T.; Magazin, N.; Petković, K.; Kovačević, D.; Lakićević, M.D. Vermicompost and Vermicompost Leachate Application in Strawberry Production: Impact on Yield and Fruit Quality. *Horticulturae* **2023**, *9*, 337. https://doi.org/10.3390/horticulturae9030337

Academic Editor: Toktam Taghavi

Received: 18 January 2023
Revised: 25 February 2023
Accepted: 26 February 2023
Published: 3 March 2023

1. Introduction

Vermicomposting represents an ecologically acceptable technology for converting organic waste into nutritious compost [1]. Vermicompost contains high concentrations of nutrients in available forms that plants absorb readily [2]. At the same time, vermicompost application can be an effective means of improving soil fertility due to the positive impact on microbiological activity and soil physical properties [3]. So far, vermicompost application has shown a positive effect on the growth of various plant species such as tomato [4], pepper [5], corn [6], strawberries [7], chickpeas [8], and many others. In addition to the solid phase, as the final product, the vermicomposting process produces leachate which can be used as a liquid fertilizer because it contains a certain amount of nutrients [9,10]. Besides necessary nutrients, vermicompost leachates also aid in the development of plants because they contain substances that control plant growth, such as humic acids, auxins (0.55–0.77 pmol mL^{-1}), gibberellins (552–656 pg mL^{-1}), and cytokinins (30–340 fmol mL^{-1}) [11,12], which control a variety of processes related to plant growth and yield, including the absorption of macro- and micronutrients. Additionally, according to several authors, the application of vermicompost improves the fruit's nutritional qualities [6,13,14]. Other studies support the positive effects of the application of vermicompost leachates on the increased resistance to salt stress of white stonecrop (*Sedum album*) and pomegranate (*Punica granatum*) seedlings [15,16] as well as the increased growth of Chinese cabbage (*Brassica pekinensis*) and tomato (*Lycopersicon esculentum*) [10,17].

The strawberry (*Fragaria ananassa* Dush.) is a hugely popular and economically important fruit species in many countries, including Serbia. In 2020, the value of strawberry output globally was USD 14 billion [18]. The strawberry is a cosmopolitan plant that can be grown on different soil types and altitudes. All over the world, the area under strawberry cultivation has increased significantly in the last decade, mainly due to the increased demand [19]. Strawberry fruits are a great source of nutrients, and several bioactive substances that may be beneficial to human health. Strawberry fruits may contain antioxidant, anticancer, anti-inflammatory, and antineurodegenerative biological activities because of their high polyphenol content, particularly anthocyanins [20]. These properties of strawberry fruits can be greatly influenced by agronomical practices, especially fertilization [21].

The objective of this research was to compare and quantify the impact of vermicompost and vermicompost leachate application on the yield and quality of strawberries concerning the mineral NPK fertilization during a 3-year production cycle.

2. Materials and Methods

2.1. Experimental Site

The field experiment was conducted during 2009–2012 at a location near Novi Sad, Serbia: (45°20′24.44″ N, 19°50′22.32″ E). The experiment was conducted on clay loam soil classified as Phaeozem (the dominant soil type in northern Serbia). The soil was slightly alkaline (7.92 pH in H_2O) with a low content of calcium carbonate (0.83%) and organic matter (2.05%). The content of available phosphorus was low (24.9 mg kg^{-1} P), while the content of available potassium was at an optimal level (179 mg kg^{-1} K). The concentrations of the available form of microelements were as follows: 2.06 mg kg^{-1} DTPA-Fe; 18.56 mg kg^{-1} DTPA-Mn; 0.89 mg kg^{-1} DTPA-Cu; 1.26 mg kg^{-1} DTPA-Zn.

The experimental plot was equipped with a drip irrigation system with water from an artesian well. Irrigation was carried out every year from April to September when the irrigation system was automatically switched on to keep the soil moisture tension within the range of 15 to 25 kPa. A tensiometer was used to monitor soil moisture 15 cm below the surface in the space between two strawberry plants. The chemical properties of irrigation water were as follows: dry residue 431 mg L^{-1}; pH 7.31; EC 0.71 dS m^{-1}; HCO_3^- 6.83 meq L^{-1}; Cl^- 3.46 meq L^{-1}; SO_4^{-2} 2.03 meq L^{-1}; Ca^{2+} 2.11 meq L^{-1}; Mg^{2+} 3.02 meq L^{-1}; Na^+ 3.98 meq L^{-1}; K^+ 0.07 meq L^{-1}; Fe 1.1 mg L^{-1}; Mn 0.11 mg L^{-1}; Cu < 0.01 mg L^{-1}; Zn 0.026 mg L^{-1}.

The average air temperature and precipitation data in the study years were obtained from on-site meteorological stations and are presented in Figure 1.

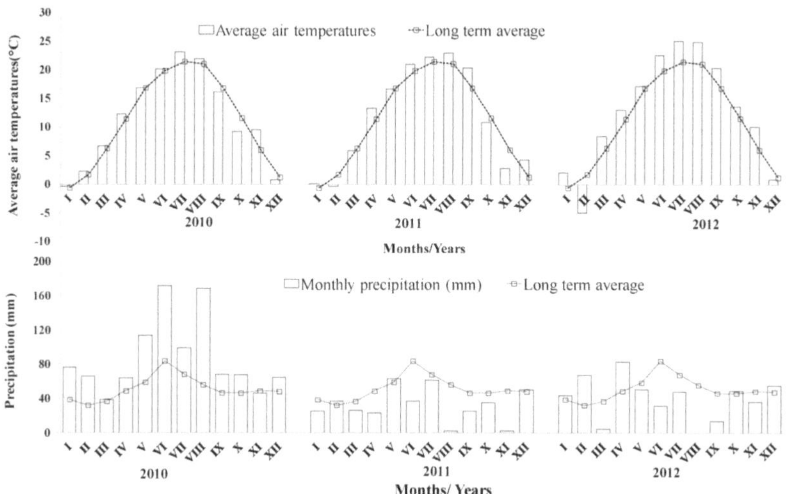

Figure 1. Average air temperature and total monthly precipitation for hydrological years (2010–2012).

2.2. Experiment Design

The experiment included five treatments arranged as completely randomized block designs in four replications. Each individual trial plot consisted of 10 strawberry plants of the June-bearing cultivar Senga Sengana, and the experiment consisted of a total of 20 plots. The experiment was set up on a plot where mineral fertilizers had not been applied for the past 3 years, and the crop that preceded the planting of strawberries was winter wheat. After the wheat harvest in mid-July 2009, the harvest residues were removed from the plot, and the soil was plowed and prepared for planting strawberry seedlings.

A week before strawberry planting, raised beds 15 cm high and 80 cm wide were formed with two drip irrigation hoses with a dropper capacity of $2\,dm^{-3}\,h^{-1}$. After that, 50 μm thick and 1.3 m wide black polyethylene perforated foil (10 plants per m^2) was stretched over the raised beds. Strawberry planting was performed manually on 23 July 2009. During the entire period of the experiment, fungicides were applied every year to prevent the most prevalent strawberry diseases: common leaf spot (*Mycosphaerella fragariae*) and fruit rot (*Botrytis cinerea*).

The experiment consisted of five fertilization treatments:

1. Control treatment—no fertilization (Ø).
2. Preplant application of vermicompost in the amount which adds $170\,kg\,N\,ha^{-1}$ (V) to the soil.
3. Preplant application of vermicompost ($170\,kg\,N\,ha^{-1}$) and foliar application of vermicompost leachate during vegetation (four times during April–May and three times in 15-day intervals during July–August (seven sprays per year in total) (VL)).
4. Foliar application and fertigation with vermicompost leachate (four times during April–May and three times in 15-day intervals during July–August (seven sprays per year in total) (L)).
5. Standard fertilization with mineral fertilizers (NPK). A total of $60\,kg\,N\,ha^{-1}$, $30\,kg\,P_2O_5\,ha^{-1}$, and $80\,kg\,K_2O\,ha^{-1}$ were applied each year as ammonium nitrate, superphosphate, and potassium nitrate, respectively.

In treatments 2 and 3 (V and VL), the vermicompost was incorporated into the surface soil layer (0–30 cm) 1 week before strawberry planting (23 July 2009). At the same time, in treatment 5 (NPK treatment), $300\,kg\,ha^{-1}$ of compound mineral fertilizer 8:16:24 and 100 kg of ammonium nitrate were applied. Additionally, in this treatment, as a standard fertilization practice, mineral NPK fertilizers were applied through a drip irrigation system once a week from April to mid-September in all three growing seasons. The total applied amounts of nitrogen, phosphorus, and potassium through different fertilization treatments, as well as the time and method of fertilizer application, are shown in Table 1.

Table 1. Total amounts of nitrogen, phosphorus, and potassium applied through different fertilization treatments, time, and method of fertilizer application.

Fertilization Treatments	Time and Method of Fertilizer Application								
	Before Planting (Year 2009) (Plowed)			During Vegetation (Years 2009–12) (Fertigation)			During Vegetation (Years 2009–12) (Foliar Application)		
	N $(kg\,ha^{-1})$	P_2O_5 $(kg\,ha^{-1})$	K_2O $(kg\,ha^{-1})$	N $(kg\,ha^{-1})$	P_2O_5 $(kg\,ha^{-1})$	K_2O $(kg\,ha^{-1})$	N $(kg\,ha^{-1})$	P_2O_5 $(kg\,ha^{-1})$	K_2O $(kg\,ha^{-1})$
Ø	-	-	-	-	-	-	-	-	-
V	170	251	110	-	-	-	-	-	-
V + L	170	251	110	-	-	-	2.5	4.47	6.3
L	-	-	-	25	44.5	63	2.5	4.47	6.3
NPK	60	80	120	70	40	80	-	-	-

Ø, control; V, vermicompost; V + L, vermicompost + leachate; L, leachate; NPK, mineral fertilizer.

The application of mineral fertilizers was carried out under the same conditions during the growing season when vermicompost leachate was applied, with two-thirds of the total amount of fertilizer applied in the first half of the growing season (before the strawberry blossoms), while the rest was applied in the second half of the growing season during August and September each year.

The vermicompost leachate used in this research was collected on the same farm where the vermicompost was produced. The required amount of vermicompost leachate used during the growing season was collected every year in the spring. Every year before application, the samples of vermicompost leachate were analyzed, and the average values of the chemical composition of three samples (for three years of application) are shown in Table 2. During the growing season, vermicompost leachate was stored in a refrigerator at a temperature of 4 °C.

Table 2. Chemical properties of organic fertilizers.

Chemical Properties	Organic Fertilizers	
	Vermicompost	Vermicompost Leachate
Bulk density (g cm^{-3})	-	1.05
Dry mater (g kg^{-1})	754	12.6 ± 0.64
pH	7.56	7.21 ± 0.11
Total N (g kg^{-1})	19.9	0.357 ± 0.05
NO$_3$-N (mg kg^{-1})	450	306.2 ± 47.2
NH$_4$-N (mg kg^{-1})	86.1	50.5 ± 11.15
Organic C (g kg^{-1})	274	17.5 ± 1.94
C/N ratio	13.8	49.2 ± 2.30
Total P (g kg^{-1})	13.2	0.28 ± 0.06
Water soluble P (mg kg^{-1})	621	0.028 ± 0.0006
Total K (g kg^{-1})	10.5	0.75 ± 0.14
Water soluble K (mg kg^{-1})	798	0.075 ± 0.22
Total Ca (g kg^{-1})	18.6	0.102 ± 0.02
Total Mg (g kg^{-1})	6.50	0.074 ± 0.015
Fe (mg kg^{-1})	1054	96 ± 13.05
Mn (mg kg^{-1})	171	1.35 ± 0.91
Cu (mg kg^{-1})	8.90	2.5 ± 0.86
Zn (mg kg^{-1})	45.2	3.0 ± 0.90

For fertigation, aqueous vermicompost extracts were prepared by diluting vermicompost leachate with tap water at a ratio of 1:2 (vol/vol). Before application, the solution was filtered through filter paper and applied at the rate of 2 L m^{-2}. Foliar application of vermicompost leachate at the rate of 0.1 L m^{-2} (filtered undiluted solution) was performed with a hand sprayer in all three growing seasons. Fertigation with vermicompost leachate was performed at the same time as the foliar application. Prior to the application of vermicompost leachate, all the plots were irrigated for an hour before the leachate application. The chemical compositions of applied vermicompost and vermicompost leachate are presented in Table 2.

2.3. Measurements and Analytical Determination

The soil pH value was measured in a soil/water suspension (1:2.5 ratio, respectively) with a Metrel MA 3657 pH meter. The calcium carbonate (CaCO$_3$) content in the soil was measured using a Scheibler calcimeter, a volumetric method. Soil organic matter (organic C) was analyzed by a CHNS analyzer (Elementar Vario EL, GmbH, Hanau, Germany).

Available-form phosphorus and potassium in the soil were analyzed after extraction with an AL solution (0.1 mol L^{-1} ammonium lactate and mol L^{-1} acetic acid, pH 3.75) at a soil-to-solution ratio of 1:20 (w/v) [22]. The concentration of available phosphorus was measured spectrophotometrically, while the concentration of K was measured by flame photometric technique. Plant-available fractions of Fe, Mn, Cu, and Zn in the soil samples were measured after extraction with DTPA-TEA buffer (0.005 mol L^{-1} DTPA + 0.01 mol L^{-1} CaCl$_2$ + 0.1 mol L^{-1} TEA)) by an atomic absorption spectrometer (Shimadzu 6300, Kyoto, Japan). The dry matter content in vermicompost was determined using the gravimetric method by drying to a constant weight at 70 °C for 24 h.

Ground samples of vermicompost were analyzed for total C and N using an automated CHNS analyzer (Elementar Vario EL, GmbH, Hanau, Germany).

The contents of K, Fe, Mn, Cu, and Zn in vermicompost and vermicompost leachate were analyzed by the wet digestion method (mixture of HNO$_3$:HClO$_4$), while the concentrations of these elements in the solution were determined by the method of atomic absorption spectrophotometry (AAS, Shimadzu 6300). The concentration of P in vermicompost and leachate was measured by the spectrophotometric molybdovanadate method after wet digestion with hydrochloric acid [23]. Mineral N in the vermicompost was extracted by 2 mol L^{-1} KCl (1:4, soil-to-solution ratio, weight basis) and determined by steam distillation [24].

The strawberry yield was calculated by taking into account all harvested fruits. In order to determine the average weight of the fruit, 30 strawberry fruits were randomly selected from each repetition at the peak of the harvest period and weighed separately. The quality parameters of strawberry fruit were determined from the same samples taken for average weight.

The TMS-PRO texture analyzer (Food Technology Corporation—Sterling, VA, USA) with a 5 mm diameter stainless steel probe was used for determining the strawberry fruit firmness. The firmness of the fruit represents the mean value of the resistance force of strawberry fruit from the equatorial side and is measured in newtons (N).

Strawberry fruit color was measured using the Konica Minolta CR-400 three-filter colorimeter (Mississauga, ON, Canada). The measured color values of strawberry fruit are based on the CIE $L^*a^*b^*$ color system (Commission Internationale de l'Éclairage or International Commission on Illumination, CIE), where value L^* represents the luminance (illumination, lightness), and a^* and b^* represent the color. Based on the obtained readings for the values of a^* and b^*, value C was calculated, which represents the chromaticity or purity of the color, as well as the angle h°, which defines the intensity of the red color (0° = purple red, 90° = yellow, 180° = bluish-green, 270° = blue) [25].

Fruit slices, obtained from 30 whole berries in 4 replicates, were homogenized in a blender and used to determine total soluble solids (TSS) and titratable acidity (TA).

The concentration of TSS (expressed in Brix°) in the fresh mass of strawberry fruit was measured by direct reading using a hand-held refractometer (Hunan Xiangxin instruments, Changsha, China), whereas TA was measured on filtered strawberry juice by the titration method with 0.1 M NaOH (the equivalence point was at pH 8.1). The volume of NaOH solution used for titration was multiplied by the correction factor (0.64), and the results were expressed as percentages.

The concentration of total anthocyanins in the fresh fruits of strawberries was determined by the pH differential method with two buffer systems as buffer solutions: potassium chloride, pH 1 (0.025 mol L^{-1}), and sodium acetate, pH 4.5 (0.4 mol L^{-1}) [26]. The anthocyanin content in the fresh strawberry fruit mass was expressed in mg of pelargonidin-3-glucoside equivalents per 100 g of fresh strawberry fruit. In the same samples, the antioxidant activity of strawberry fruit was determined by the FRAP method (ferric reducing antioxidant power) [27]. The total antioxidative capacity of strawberry fruit was expressed in FRAP units, wherein one FRAP unit is equivalent to 100 µM Fe^{2+}.

To determine the mineral composition, thirty whole berries randomly selected from each replicate were mixed thoroughly and then dried at 80 °C for 72 h to the constant weight in a forced air oven. After that, the mineral composition was determined by the method of wet digestion with nitric acid (HNO_3) in a microwave oven at a temperature of 200 °C and a pressure of 170 psi, while the concentrations of K, Ca, Mg, Fe, Mn, Cu, and Zn in the obtained solution were measured by atomic absorption spectrophotometry (Shimadzu 6300).

2.4. Statistics

All data obtained by measurements in the field experiment were subjected to analysis of variance (ANOVA), and the significance of differences between treatment means was assessed using Tukey's test (probability level, $p < 0.05$). The statistical analyses were performed using the STATISTICA 9.0 software (StatSoft Inc., Tulsa, OK, USA).

3. Results

3.1. Strawberry Yield

The preplant application of vermicompost (treatments V and V + L) and standard fertilization with mineral fertilizers (NPK) had a positive effect on the total yield of strawberries in the first year of fruiting (Figure 2). In these treatments, the strawberry yield ranged between 902 and 947 g plant^{-1} and was significantly higher than the yield achieved on the control (737 g plant^{-1}) and L treatment (751 g plant^{-1}) (Figure 2). However, in the second and third years of fruiting, a significantly higher strawberry yield compared to control was recorded only in the NPK treatment, whereas the residual effects of vermicompost application (V and V + L treatments) were not detected in the second and third fruiting years of strawberries (Figure 2). In the first fruiting year, the number of flowers per strawberry plant on the V, V + L, and NPK treatments was significantly higher compared to control. On the other hand, in the second and third year of fruiting, only the plants in the NPK treatment had a significantly higher number of flowers compared to the control and other fertilization treatments. In all three fruiting years, the average fruit weight did not differ significantly between treatments (Figure 2).

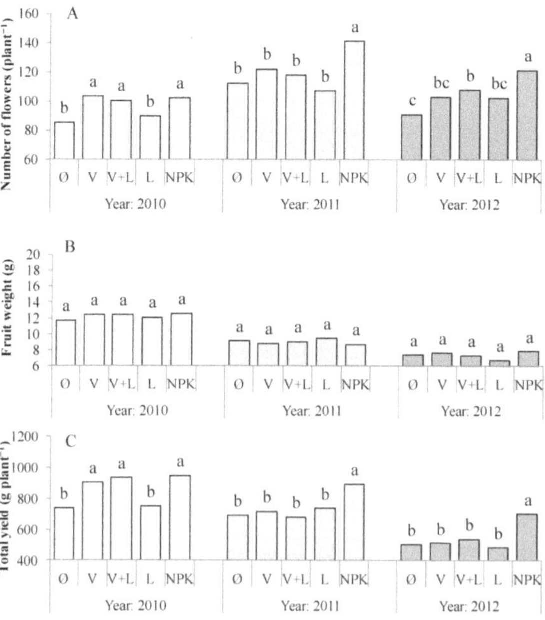

Figure 2. The total number of flowers (**A**), average strawberry fruit weight (**B**), and total yield of strawberries (**C**). Ø, control; V, vermicompost; V + L, vermicompost + leachate; L, leachate; NPK, mineral fertilizer. Different letters denote a significant difference at $p < 0.05$ for each year separately.

3.2. Mineral Composition of Strawberry Fruit

The mineral composition of strawberry fruit over three years of fruiting is shown in Table 3. In the first year of fruiting, in all four fertilization treatments, the concentration of K in strawberry fruit was significantly higher compared to control. In the second and third years of fruiting, a higher concentration of K compared to control was measured only in treatments L and NPK. On the other hand, the application of vermicompost and leachate did not lead to a significant increase in the concentration of the other elements. In contrast, standard fertilization with NPK fertilizers led to increased concentrations of K, Fe, Mn, Cu, and Zn in strawberry fruit in all three years of fruiting.

Table 3. Mineral composition of strawberry fruit.

Fertilization	K g kg^{-1}	Ca g kg^{-1}	Mg g kg^{-1}	Fe mg kg^{-1}	Mn mg kg^{-1}	Cu mg kg^{-1}	Zn mg kg^{-1}
2010 (First Fruiting Year)							
Ø [1]	1.11 c	125 a	102 a	3.98 b	3.72 b	0.39 a	0.80 c
V	1.26 b	120 a	102 a	5.13 ab	3.99 b	0.38 a	0.83 bc
V + L	1.27 b	117 a	96 a	5.75 ab	3.79 b	0.37 a	0.95 b
L	1.42 a	119 a	105 a	4.47 ab	3.95 b	0.38 a	0.74 c
NPK	1.37 ab	114 a	112 a	6.99 a	4.98 a	0.42 a	1.19 a
2011 (Second Fruiting Year)							
Ø	1.32 c	149 a	125 a	6.60 b	5.82 b	0.70 b	1.21 b
V	1.34 bc	152 a	130 a	6.17 b	6.22 b	0.79 b	1.23 b
V + L	1.36 b	155 a	126 a	5.44 b	6.35 b	0.65 bc	1.21 b
L	1.60 a	135 a	129 a	6.35 b	6.06 b	0.52 c	1.25 b
NPK	1.53 ab	146 a	137 a	10.5 a	9.08 a	0.94 a	1.35 a
2012 (Third Fruiting Year)							
Ø	1.56 b	171 a	148 ab	3.61 b	3.37 b	0.34 b	0.57 b
V	1.59 b	155 a	136 b	3.44 b	3.45 b	0.22 c	0.74 b
V + L	1.42 b	157 a	133 b	3.27 b	2.27 c	0.28 bc	0.75 b
L	1.92 a	154 a	143 ab	3.32 b	3.33 b	0.31 bc	0.66 b
NPK	1.86 a	152 a	157 a	6.90 a	5.25 a	0.51 a	1.02 a

Ø, control; V, vermicompost; V + L, vermicompost + leachate; L, leachate; NPK, mineral fertilizer. Different letters denote a significant difference at $p < 0.05$ for each year separately.

3.3. Strawberry Color and Mechanical Properties

Fertilization treatments had a significant effect on fruit color parameters in all three years of fruiting (Table 4). In the first 2 years of fruiting, the lowest values of h° were measured in treatment V. In addition to a more intense color, strawberry fruits in this treatment were characterized by lower chromaticity (lower C values) compared to NPK treatment and control. The application of NPK fertilizers had the opposite effect from the application of leachate. Strawberry fruits in this treatment were characterized by lighter (higher L values) and more chromatic colors (higher C values) but less redness (higher h° values) compared to treatment L. The application of vermicompost (V and V + L treatments) did not significantly affect the color of the strawberry fruit (Table 4).

Figure 3 shows the parameters of strawberry fruit quality depending on the fertilization treatment during the three years of fruiting. The application of vermicompost leachate led to a higher TSS concentration and a higher TSS/TA ratio compared to control. Additionally, leachate application led to a significantly higher concentration of anthocyanins and antioxidant activity of the fruit (FRAP) not only in the control but also in the NPK treatment. On the other hand, the lowest firmness of strawberry fruit in all three years was measured on the treatment where vermicompost leachate was applied (L treatment). Strawberry fruit quality parameters in treatment V, in all three years of fruiting, did not differ significantly compared to the control treatment without fertilization.

Table 4. Strawberry fruit color parameters (L, lightness; C, chromaticity; h°, intensity of red color).

	2010		
Fertilization	L	C	h°
Ø	29.6 a	35.5 a	27.7 ab
V	29.5 a	32.3 ab	29.4 ab
V + L	30.6 a	34.8 a	29.2 ab
L	28.9 a	30.9 b	26.8 b
NPK	30.7 a	35.4 a	29.9 a
	2011		
	L	C	h°
Ø	30.4 a	29.0 a	29.8 ab
V	29.7 ab	27.9 ab	32.8 a
V + L	30.6 a	28.4 a	31.0 ab
L	28.6 b	25.3 b	29.7 b
NPK	31.2 a	30.0 a	33.1 a
	2012		
	L	C	h°
Ø	33.3 ab	38.2 b	33.5 b
V	33.4 ab	37.1 bc	33.6 b
V + L	31.9 b	35.0 cd	34.5 b
L	32.3 b	33.4 d	33.7 b
NPK	34.7 a	41.5 a	37.3 a

Ø, control; V, vermicompost; V + L, vermicompost + leachate; L, leachate; NPK, mineral fertilizer. Different letters denote a significant difference at $p < 0.05$ for each year separately.

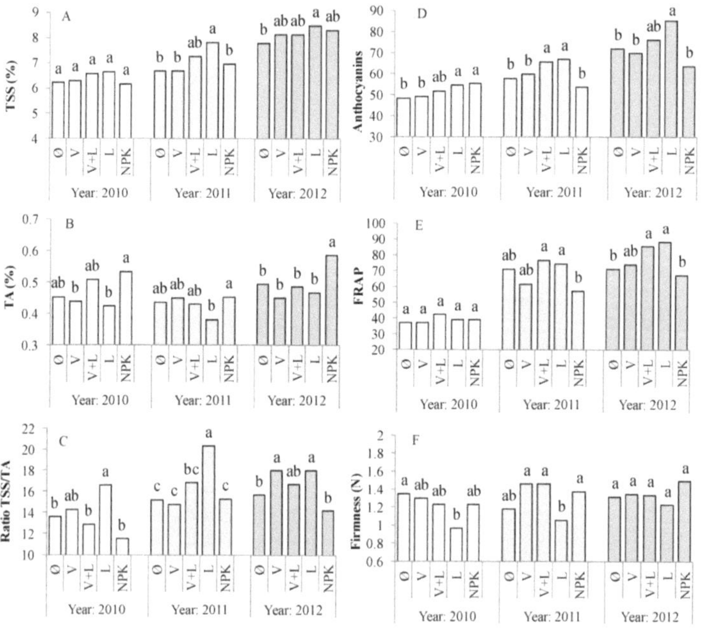

Figure 3. The concentration of total soluble solids (TSS) (**A**), total acids (TA) (**B**) and the ratio between TSS and TA (**C**), anthocyanins (mg cyanidin-3-glucoside/100 g of fruit) (**D**), antioxidant activity (FRAP units) (**E**), and the firmness of strawberry fruit (**F**). Ø, control; V, vermicompost; V + L, vermicompost + leachate; L, leachate; NPK, mineral fertilizer. Different letters denote a significant difference at $p < 0.05$ for each year separately.

4. Discussion

The treatments where vermicompost was applied at the time of planting (treatments V and V + L) and the treatment with a standard fertilization program with NPK fertilizers led to a significant increase in strawberry yield in the first year of fruiting. In these treatments, the yield of strawberries ranged between 813 and 829 g plant^{-1}, respectively, and was significantly higher than the control (693 g plant^{-1}) and the treatment where vermicompost leachate was applied (705 g plant^{-1}) (Figure 2). Such results are in agreement with the previous research by [25,28], who also reported the positive effect of the preplant application of vermicompost on strawberry yield. However, in the second and third years of fruiting, a significantly higher strawberry yield compared to control was observed only in the NPK treatment, while the residual effects of vermicompost application were not registered in the second and third years of fruiting (Figure 2).

In our study, vermicompost was applied prior to strawberry planting. Due to favorable conditions for mineralization (favorable temperatures for mineralization during August), most of the nitrogen from vermicompost likely became available to the strawberry plants in the year of application. In this case, the release of nutrients and primarily nitrogen from the vermicompost coincided with the period of flower differentiation. It is possible that the plants formed a higher reserve of N, leading to more flowers per plant in the first year of fruiting (Figure 2), which in turn led to an increase in yield due to a higher number of fruits per strawberry plant [29,30].

The application of vermicompost leachate in our study did not increase strawberry yield, which is in contrast to the studies of [31,32], who reported the positive effect of foliar application (seven times during vegetation) of compost extract on strawberry yield. The reason for the absence of a positive influence of vermicompost leachate on the yield of strawberries in our study may be the relatively high fertility of the soil on which the experiment was conducted, and the relatively low nitrogen content in vermicompost leachate (Table 2).

The application of vermicompost significantly increased the concentration of K and Zn in strawberry fruit only in the first year of fruiting, while the fertigation and foliar application of vermicompost leachate (L treatment) significantly increased the concentration of K compared to control during all three years of fruiting.

On the other hand, in the NPK treatment, where fertilization was performed during all three years of fruiting, the concentrations of K, Fe, Mn, and Zn in strawberry fruit were significantly higher compared to control. These results are consistent with the research of [33,34], who also reported the positive influence of fertilization with macroelements, primarily nitrogen, on the uptake of microelements.

The concentration of TSS in strawberry fruit, in all three years of fruiting, had values that ranged from 6.01% to 8.47%. In the first year, no significant differences were registered between treatments regarding TSS concentration in strawberry fruit, despite a significant difference in yield between treatments. In the first year of fruiting, significantly more precipitation was measured not only compared to the second and third years of fruiting but also compared to the long-term average (Figure 1). Heavy precipitation, combined with cloudy weather, may be the reason why the concentration of TSS in strawberry fruit did not differ significantly between treatments. Ref. [35] also reported the enormous influence of weather conditions during strawberry harvest on the concentration of TSS and the chemical composition of strawberry fruit in general. In the second and third years of fruiting, the highest TSS concentration was measured on treatment L, which led to a significantly higher TSS/TA ratio compared to other treatments.

The authors of [31] also reported a higher TSS concentration and lower TA values in strawberry fruit compared to the control treatment due to the application of vermicompost leachate. According to [36], the foliar application of a vermicompost extract led to an increase or decrease in fruit firmness depending on the tomato variety, while in all varieties, it led to a decrease in the concentration of ascorbic acid. On the other hand, the highest TA was measured in the fruits on NPK treatment. The increased acidity of strawberry fruit may be a consequence of the higher amount of N applied on this treatment (compared to other treatments), which could lead to higher fruit acidity and a lower TSS/TA ratio [21,37].

The highest anthocyanin concentration and antioxidant activity of the fruit (FRAP values) were measured on the treatment where vermicompost leachate was applied. The vermicompost leachate used in our study had a relatively high K concentration (Table 2), which may have affected the anthocyanin concentration in strawberry fruit [38]. Additionally, vermicompost leachate contains phytohormones [11,12], which could have a significant role in increasing the concentration of anthocyanins in strawberry fruit, especially gibberellic acid [39].

The higher concentration of anthocyanins due to the application of vermicompost leachate resulted in a change in the color of strawberry fruit [40], so the fruits had a more intense red color (lower values of h°) and a darker and less chromatic color compared to the treatment with NPK fertilizers, and in some years, compared to the control as well.

Additionally, the higher concentrations of TSS and anthocyanins and the darker fruit color all indicate a higher degree of fruit maturity, which explains why the fruits had the lowest fruit firmness after the application of vermicompost leachate [41].

The influence of fertilization treatment on certain parameters of strawberry quality differed between the years of fruiting, which indicates the existence of interactions between external factors (temperature, precipitation, etc.) and applied fertilizers [37].

5. Conclusions

The results showed that the preplant application of vermicompost applied in a relatively small dose (equivalent to 170 kg N/ha) had a positive effect on the yield, which in the first year of fruiting was comparable to the yield obtained in the treatment with the standard application of NPK fertilizers. On the other hand, in the second and third years of fruiting, the yield on this treatment was at the level of the control treatment (without fertilization) and significantly lower compared to the treatment with standard fertilization with NPK fertilizers.

The application of vermicompost leachate by fertigation and foliar application did not affect strawberry yield but had a positive effect on strawberry fruit quality parameters such as total soluble solids, total anthocyanins, and antioxidant activity of strawberry fruit. If strawberries are grown for 2 or 3 years in the same place, the preplant application of vermicompost and vermicompost leachate application during vegetation cannot maintain the yield level as with the application of mineral NPK fertilizers, but the quality of strawberry fruit can be improved significantly.

Recycling of organic wastes is becoming increasingly important as the need to protect natural resources grows, and vermicomposting as a technology can certainly make a significant contribution. Further research should provide answers to the economic aspects of the application of vermicompost and vermicompost leachate in strawberry production as well as answers to the eventual contamination of strawberry fruit with harmful substances that can be found in organic fertilizers.

Author Contributions: Conceptualization, R.Č., M.S.M. and K.P.; methodology, R.Č., B.M.P., M.T.R. and N.M.; software, M.D.L.; validation, R.Č. and N.M.; writing—original draft preparation, R.Č. and K.P.; writing—review and editing, D.K. and M.D.L.; supervision, M.S.M.; funding acquisition, M.S.M. All authors have read and agreed to the published version of the manuscript.

Funding: This work was supported by the Ministry of Science, Technological Development and Innovation of the Republic of Serbia (no. 451-03-47/2023-01/200117).

Data Availability Statement: Not applicable.

Conflicts of Interest: The authors declare no conflict of interest.

References

1. Orozco, S.H.; Cegarra, J.; Trujillo, L.M.; Roig, A. Vermicomposting of coffee pulp using the earthworm *Eisenia fetida*: Effects on C and N contents and the availability of nutrients. *Biol. Fert. Soils* **1996**, *22*, 162–166. [CrossRef]
2. Blouin, M.; Barrere, J.; Meyer, N.; Lartigueet, S.; Barot, S.; Mathieu, J. Vermicompost significantly affects plant growth. A meta-analysis. *Agron. Sustain. Dev.* **2019**, *39*, 34. [CrossRef]
3. Lim, S.L.; Wu, T.Y.; Lim, P.N.; Shak, K.P.Y. The use of vermicompost in organic farming: Overview, effects on soil and economics. *J. Sci. Food Agric.* **2015**, *95*, 1156–3143. [CrossRef] [PubMed]
4. Goswami, L.; Nath, A.; Sutradhar, S.; Bhattacharya, S.S.; Kalamdhad, A.; Vellingiri, K.; Kim, K.H. Application of drum compost and vermicompost to improve soil health, growth, and yield parameters for tomato and cabbage plants. *J. Environ. Manag.* **2017**, *200*, 243–252. [CrossRef]
5. Castellanos, J.Z.; Cano-Ríos, P.; García-Carrillo, E.M.; Olalde-Portugal, V.; Preciado-Rangel, P.; Ríos-Plaza, J.L.; García-Hernández, J.L. Hot pepper (*Capsicum annuum* L.) growth, fruit yield, and quality using organic sources of nutrients. *Compost Sci. Util.* **2017**, *25*, S70–S77. [CrossRef]
6. Lazcano, C.; Revilla, P.; Malvar, R.A.; Domínguez, J. Yield and fruit quality of four sweet corn hybrids (*Zea mays*) under conventional and integrated fertilization with vermicompost. *J. Sci. Food Agric.* **2011**, *91*, 1244–1253. [CrossRef]
7. Zuo, Y.; Zhang, J.; Zhao, R.; Dai, H.; Zhang, Z. Application of vermicompost improves strawberry growth and quality through increased photosynthesis rate, free radical scavenging and soil enzymatic activity. *Sci. Hortic.* **2018**, *233*, 132–140. [CrossRef]
8. Hosseinzadeh, S.R.; Amiri, H.; Ismaili, A. Evaluation of photosynthesis, physiological, and biochemical responses of chickpea (*Cicer arietinum* L. cv. Pirouz) under water deficit stress and use of vermicompost fertilizer. *J. Integr. Agric.* **2018**, *17*, 2426–2437. [CrossRef]
9. Gutiérrez-Miceli, F.A.; García-Gómez, R.C.; Oliva-Llaven, M.A.; Montes-Molina, J.A.; Dendooven, L. Vermicomposting leachate as liquid fertilizer for the cultivation of sugarcane (*Saccharum* sp.). *J. Plant Nutr.* **2017**, *40*, 40–49. [CrossRef]
10. Churilova, E.V.; Midmore, D.J. Vermiliquer (vermicompost leachate) as a complete liquid fertilizer for hydroponically-grown pak choi (*Brassica chinensis* L.) in the tropics. *Horticulturae* **2019**, *5*, 26. [CrossRef]
11. Aremu, A.O.; Stirk, W.A.; Kulkarni, M.G.; Tarkowska, D.; Tureckova, V.; Gruz, J.; Subrtova, M.; Pencik, A.; Novak, O.; Dolezal, K.; et al. Evidence of phytohormones and phenolic acids variability in garden-waste-derived vermicompost leachate, a well-known plant growth stimulant. *Plant Growth Regul.* **2015**, *75*, 483–492. [CrossRef]
12. Scaglia, B.; Nunes, R.R.; Rezende, M.O.O.; Tambone, F.; Adani, F. Investigating organic molecules responsible of auxin-like activity of humic acid fraction extracted from vermicompost. *Sci. Total Environ.* **2016**, *562*, 289–295. [CrossRef] [PubMed]
13. Gutiérrez-Miceli, F.A.; Santiago-Borraz, J.; Montes Molina, J.A.; Nafate, C.C.; Abdud-Archila, M.; Oliva Llaven, M.A.; Rincón-Rosales, R.; Deendoven, L. Vermicompost as a soil supplement to improve growth, yield and fruit quality of tomato (Lycopersicum esculentum). *Bioresour. Technol.* **2007**, *98*, 2781–2786. [CrossRef] [PubMed]
14. Wang, D.; Shi, Q.; Wang, X.; Wei, M.; Hu, J.; Liu, J.; Yang, F. Influence of cow manure vermicompost on the growth, metabolite contents, and antioxidant activities of Chinese cabbage (*Brassica campestris* ssp. chinensis). *Biol. Fertil. Soils* **2010**, *46*, 689–696. [CrossRef]
15. Cicek, N.; Erdogan, M.; Yucedag, C.; Cetin, M. Improving the Detrimental Aspects of Salinity in Salinized Soils of Arid and Semi-arid Areas for Effects of Vermicompost Leachate on Salt Stress in Seedlings. *Water Air Soil Pollut.* **2022**, *233*, 197. [CrossRef]
16. Bidabadi, S.S.; Dehghanipoodeh, S.; Wright, G.C. Vermicompost leachate reduces some negative effects of salt stress in pomegranate. *Int. J. Recycl. Org. Waste Agric.* **2017**, *6*, 255–263. [CrossRef]
17. Arthur, G.D.; Aremu, A.O.; Kulkarni, M.G.; Staden, J.V. Vermicompost leachate alleviates deficiency of phosphorus and potassium in tomato seedlings. *HortScience* **2012**, *47*, 1304–1307. [CrossRef]
18. Food and Agriculture Organization of the United Nations (FAOSTAT). *Crops and Livestock Products: Strawberries*; Food and Agriculture Organization of the United Nations: Rome, Italy, 2021.
19. Hernández-Martínez, N.; Blanchard, C.; Wells, D.; Salazar-Gutiérrez, M. Current state and future perspectives of commercial strawberry production: A review. *Sci. Hortic.* **2023**, *312*, 111893. [CrossRef]
20. Afrin, S.; Gasparrini, M.; Forbes-Hernandez, T.Y.; Reboredo-Rodriguez, P.; Mezzetti, B.; Varela-López, A.; Giampieri, F.; Battino, M. Promising Health Benefits of the Strawberry: A Focus on Clinical Studies. *J. Agric. Food Chem.* **2016**, *64*, 4435–4449. [CrossRef]
21. Ojeda-Real, L.A.; Lobit, P.; Cardenas-Navarro, R.; Grageda-Cabrera, O.; Farias-Rodriguez, R.; Valencia-Cantero, E.; Macias-Rodriguez, L. Effect of nitrogen fertilization on quality markers of strawberry (*Fragaria* × *ananassa* Duch. cv. Aromas). *J. Sci. Food Agric* **2009**, *89*, 935–939. [CrossRef]
22. Egner, H.; Riehm, H.; Domingo, W.R. Investigations of the chemical soil analysis as a basis for the evaluation of nutrient status in soil II: Chemical extraction methods for phosphorus and potassium determination. *K Lantbr. Høgsk Ann* **1960**, *26*, 199–215.
23. Faithfull, N.T. *Methods in Agricultural Chemical Analysis. A Practical Handbook*; CABI Publishing: Wallingford, UK, 2002.
24. Bremner, J.M. Nitrogen availability indexes. In *Methods of Soil Analysis: Part 2 Chemical and Microbiological Properties, 9.2*; Norman, A.G., Ed.; American Society of Agronomy: Madison, WI, USA, 1965; pp. 1324–1345.
25. Singh, R.; Sharma, R.R.; Tyagi, S.K. Pre-harvest foliar application of calcium and boron influences physiological disorders, fruit yield and quality of strawberry (*Fragaria* × *ananassa* Duch.). *Sci. Hortic.* **2007**, *112*, 215–220. [CrossRef]

26. Lako, J.; Trenerry, V.C.; Wahlqvist, M.; Wattanapenpaiboon, N.; Sotheeswaran, S.; Premier, R. Phytochemical flavonols, carotenoids and the antioxidant properties of a wide selection of Fijian fruit, vegetables and other readily available foods. *Food Chem.* **2007**, *101*, 1727–1741. [CrossRef]
27. Benzie, F.F.; Strain, J.J. Ferric reducing/antioxidant power assay: Direct measure of total antioxidant activity of biological fluids and modified version for simultaneous measurement of total antioxidant power and ascorbic acid concentration. *Methods Enzymol.* **1999**, *299*, 15–27. [PubMed]
28. Beck, J.E.; Schroeder-Moreno, M.S.; Fernandez, G.E.; Grossman, J.M.; Creamer, N.G. Effects of Cover Crops, Compost, and Vermicompost on Strawberry Yields and Nitrogen Availability in North Carolina. *Horttechnology* **2016**, *26*, 604–613. [CrossRef]
29. Opstad, N.; Nes, A.; Mage, F. Preplant fertilization and fertigation in strawberry (*Fragaria* × *ananassa* Duch. cv. 'Korona') in an open field experiment. *Eur. J. Hortic. Sci.* **2007**, *72*, 206–213.
30. Nestb, R.; Guéry, S. Balanced fertigation and improved sustainability of June bearing strawberry cultivated three years in open polytunnel. *J. Berry Res.* **2017**, *7*, 203–216. [CrossRef]
31. Singh, R.; Gupta, R.K.; Patil, R.T.; Sharma, R.R.; Asrey, R.; Kumar, A.; Jangra, K.K. Sequential foliar application of vermicompost leachates improves marketablefruit yield and quality of strawberry (*Fragaria* × *ananassa* Duch.). *Sci. Hortic.* **2010**, *124*, 34–39. [CrossRef]
32. Welke, S.E. The Effect of Compost Extract on the Yield of Strawberries and the Severity of Botrytis cinerea. *J. Sustain. Agric.* **2004**, *25*, 57–68. [CrossRef]
33. Klikocka, H.; Podlesna, A.; Podlesny, J.; Narolski, B.; Haneklaus, S.; Bloem, E.; Schnug, E. Improvement of the content and uptake of micronutrients in spring rye grain DM through nitrogen and sulfur supplementation. *Agronomy* **2020**, *10*, 35. [CrossRef]
34. Gołab-Bogacz, I.; Helios, W.; Kotecki, A.; Kozak, M.; Jama-Rodzeńska, A. Effect of nitrogen fertilization on the dynamics of concentration and uptake of selected microelements in the biomass of *Miscanthus* × *giganteus*. *Agriculture* **2021**, *11*, 360. [CrossRef]
35. Kannaujia, P.K.; Asrey, R. Effect of harvesting season and cultivars on storage behaviour, nutritional quality and consumer acceptability of strawberry (Fragaria × ananassa Duch.) fruits. *Acta Physiol. Plant.* **2021**, *43*, 88. [CrossRef]
36. Zaller, J.G. Foliar spraying of vermicompost extracts: Effects on fruit quality and indications of late-blight suppression of field-grown tomatoes. *Biol. Agric. Hortic.* **2006**, *24*, 165–180. [CrossRef]
37. Agehara, S.; Nunes, M.C.D.N. Season and Nitrogen Fertilization Effects on Yield and Physicochemical Attributes of Strawberry under Subtropical Climate Conditions. *Agronomy* **2021**, *11*, 1391. [CrossRef]
38. Khayyat, M.; Tafazoli, E.; Eshghi, S.; Rahemi, M.; Rajaee, S. Salinity, supplementary calcium and potassium effects on fruit yield and quality of strawberry (Fragaria ananassa Duch.). *Am.-Eurasian J. Agric. Environ. Sci.* **2007**, *2*, 539–544.
39. Roussos, P.A.; Denaxa, N.K.; Damvakaris, T. Strawberry fruit quality attributes after application of plant growth stimulating compounds. *Sci. Hortic.* **2009**, *119*, 138–146. [CrossRef]
40. Yoshida, Y.; Koyama, N.; Tamura, H. Color and anthocyanin composition of strawberry fruit: Changes during fruit development and differences among cultivars, with special reference to the occurrence of pelargonidin-malonylglucoside. *J. Jpn. Soc. Hortic. Sci.* **2002**, *71*, 355–361. [CrossRef]
41. Rahman, M.M.; Moniruzzaman, M.; Ahmad, M.R.; Sarker, B.C.; Alam, M.K. Maturity stages affect the postharvest quality and shelf-life of fruits of strawberry genotypes growing in subtropical regions. *J. Saudi Soc. Agric. Sci.* **2016**, *15*, 28–37. [CrossRef]

 *horticulturae*

Brief Report

Impact of Water and Nutrient Supplementation on Yield of Prairie Plantings of Juneberry *Amelanchier alnifolia* Nutt., Cultivar and Windbreak Plantings

Kerry Hartman [1,*], Dilmini Alahakoon [2] and Anne Fennell [2]

[1] Environmental Science, Nueta Hidatsa Sahnish College, New Town, ND 58763, USA
[2] Agronomy, Horticulture and Plant Science Department, South Dakota State University, Brookings, SD 57007, USA; dilmini.alahakoon@jacks.sdstate.edu (D.A.); anne.fennell@sdstate.edu (A.F.)
* Correspondence: khartm@nhsc.edu; Tel.: +1-701-627-8053

Abstract: *Amelanchier alnifolia* Nutt. (Juneberry, Saskatoon berry or Serviceberry) fruit historically played an important role as fresh or dried food and as a medicinal staple in the Mandan, Hidatsa, and Arikara Tribal Nations. Natural Juneberry stands were lost during the creation of Sakakawea Reservoir on the Fort Berthold Reservation. Reintroduction of the Juneberry is important to the tribal communities. Therefore, the impact of water and fertilizer supplementation was explored in two mature Juneberry cultivar (Honeywood, Martin, and Smokey) plantings and a seedling windbreak planting. Yield was examined in three consecutive years with three treatments: (1) natural conditions (control; no additional water or fertilizer); (2) irrigation during flowering and fruit ripening period (irrigated); and (3) fertilization plus irrigation during flowering and fruit ripening period (fertilized). Yield varied from 5 to 258 g/0.03 m^{-3} across locations, treatments, and years. There was no difference in yield across locations and treatments in year one. Yield was greater in the second year than first year, but not different across locations or treatments. The fertilized treatment showed increased yield in the third year in contrast to irrigated treatment across locations. New plantings can be established more economically using seedling material and the yield increased if watered and fertilized during fruit development.

Keywords: juneberry; serviceberry; saskatoon berry; naakunaánu'; máacudabaa; mánabushaké

Citation: Hartman, K.; Alahakoon, D.; Fennell, A. Impact of Water and Nutrient Supplementation on Yield of Prairie Plantings of Juneberry *Amelanchier alnifolia* Nutt., Cultivar and Windbreak Plantings. *Horticulturae* **2023**, 9, 653. https://doi.org/10.3390/horticulturae9060653

Academic Editor: Toktam Taghavi

Received: 11 April 2023
Revised: 22 May 2023
Accepted: 25 May 2023
Published: 1 June 2023

1. Introduction

Native stands of *Amelanchier alnifolia* Nutt. (Juneberry, Saskatoon berry, or Serviceberry) in the bottom land next to the river and streambanks, draws, and coulees near the Missouri river were an important fruit source for the Mandan, Hidatsa, and Arikara Tribal Nations [1]. The fruit of this hardy shrub was a food staple, important for nutrition and medicine in the Native American diet [2,3]. The Juneberry is a small pome fruit and is typically the size of a blueberry (0.5 to 1.0 cm diameter), has many small seeds, a tender skin and is eaten fresh, steamed, mashed, or dried to a brick-like consistency for winter use [1,2,4]. Although lower in vitamin C than many berry fruits, the Juneberry is a good source of copper, manganese, magnesium, and carotenoids. Juneberry fruit is a rich source of polyphenols and has a high content of flavonoid antioxidants (anthocyanins, flavonols, procyadins, and phenolic acids) [5–11]. Historically, the fruit was used to treat diarrhea, sore eyes, and stomach and liver ailments [4,9,12]. In addition, leaves, twigs, bark and roots were used to create teas and tonics. Boiled bark was used as a disinfectant, to expel worms, and to promote discharge of the placenta after delivery [12].

Phytochemical analysis of the Juneberry fruit shows a high phenolic content and free-radical scavenging capacity [9,11,13]. The ability for crude fruit extracts to inhibit nitric oxide production in macrophage suggests a protective role against cardiovascular disease and chronic inflammation [14]. In addition, exploration of the biological activity of Juneberry extracts on cell viability indicates cell viability is maintained in the presence of free radicals [13]. Bioassays of Juneberry samples show improved glucose uptake and reduced expression of anti-inflammatory markers, suggesting an ability to mediate and reduce inflammation [7]. It is notable that the flavonoid compounds in fruit extracts can protect erythrocyte membranes against oxidation by interacting with the membrane surface [15]. Analysis of the antimicrobial, antihyperglycemic, and anti-obesity potential of the Juneberry antioxidants indicate antimicrobial activity. There is also a correlation with these fruit characteristics and its ability to inhibit pancreatic lipase, free amino acids, and antihyperglycemic activity [16]. Testing Juneberry powder in rats' diets indicated it regulates glucose and ameliorates cardiovascular and liver impacts of diet-induced metabolic syndrome [17]. Thus, the anti-inflammatory, anti-diabetic, and chemoprotective effects of this fruit emphasize the importance of improving the availability of this culturally relevant fruit to the Fort Berthold Reservation.

After the completion of Garrison Dam on the Missouri River in North Dakota, USA, by the United States Army Corps of Engineers, the Sakakawea reservoir flooded the bottom land on Fort Berthold Indian Reservation and members of the Mandan, Hidatsa, and Arikara Tribal Nations relocated to the arid plateau regions. This change also flooded the native habitat of the Juneberry and impacted the community members' ability to locate and harvest fruit from native stands that once grew along the Missouri River bottom [1]. In 2004, a project was developed to reintroduce this important fruit in the upland areas near the Sakakawea reservoir [1]. Currently, there are no standardized recommendations for Juneberry irrigation and fertilization. Therefore, the objective of this study was to evaluate whether supplemental water and/or fertilizer would improve the yield in an effort to develop baseline information for future detailed plant fertilization studies in this nutritious and culturally relevant fruit for the prairie regions.

2. Materials and Methods

The Juneberry irrigation and fertilization trial was conducted from 2019 to 2021 at three locations in North Dakota, USA: White Shield, War Coulee, and Elbowoods. All sites were located on the east side of the Sakakawea reservoir in North Dakota. The White Shield cultivar planting was located in McLean County, ND. The site was a glacial till moraine with a slope of $3 \pm 1\%$, soil pH 6.8, and 6.3/15/398 ppm (nitrogen/phosphorus/potassium) and 2–4% organic matter as determined by WARD Laboratories Inc. Kearney, NE, USA. This site had wind protection due to the typography of the site. The War Coulee cultivar planting was located in McLean County, ND. This site was a glacial till moraine with a slope of $3 \pm 1\%$ near the crest, soil pH 6.0, and 6/39/160 ppm (nitrogen/phosphorus/potassium) and 3–5% organic matter as determined by WARD Laboratories Inc. Kearney, NE, USA. The site was more exposed to wind than White Shield due to its elevated location. The Elbowoods seedling windbreak plot was in Mountrail County, ND, USA. The Elbowoods site was located on a glacial till moraine with a slope of $22 \pm 1\%$, soil pH 8.0, (nitrogen/phosphorus/potassium) and 5% organic matter as determined by WARD Laboratories Inc. Kearney, NE, USA.

White Shield and War Coulee plantings contained replicated blocks of clonally propagated Juneberry cultivars 'Martin', 'Honeywood', and 'Smokey' planted in 2004 [1]. The cultivars were replicated in groups of 30 or 33 bushes per cultivar and the three groups randomized in each of the four rows [1]. The plants were spaced at 0.9 m within rows and the rows were 3.66 m apart. Bushes at White Shield had a height and spread of 2.4 m × 0.9 m and War Coulee had a height and spread of 2.0 m × 0.9 m. The Elbowoods plot was established in 2016 with seedlings obtained from the Towner State Nursery (NDSU-North Dakota Forest Service, Towner, ND, USA). The seed was collected from fruit collected from wild stands, germinated, grown for one season and dormant bare root plants received from the nursery were planted in 2016. The Elbowoods planting planted in 2016 was composed of six rows (30 plants each), with plants spaced 0.9 m within the row and 3.66 m between rows. This planting was just beginning to bear fruit in 2019 and the height and spread of the bushes was 1.0 m × 0.5 m.

Three treatments were applied to ten consecutive plants/cultivar within each of the four rows: (1) no irrigation (control); (2) irrigation only (every third day at 18.93 l per bush using drip irrigation); (3) irrigation and fertilization (irrigation every third day at 18.93 l per bush and fertilization with the irrigation once per month. The irrigation treatment was applied every third day from April 15 to September 15 and the fertilizer treatment applied once per month at experimental sites in 2019, 2020, and 2021. A general-purpose fertilizer (Jack's 20-20-20) was delivered through the irrigation water via trickle irrigation, with each bush receiving 880 ppm N (46.5 ppm each liter) on or around the 15th of the month. All irrigation water was trucked to experimental sites, which were five to six miles from the Sakakawea Reservoir.

Harvest was completely random across the treatments and was conducted when Juneberry fruit reached full color on July 10th, 1st, and 11th in 2019, 2020, and 2021, respectively. Yield (g/0.03 m^{-3}) collected once per bush for seven randomly selected bushes in each treatment from across the 4 rows. A 0.03 m^3 cube was constructed using 2.5 cm diameter polyvinyl chloride plastic pipe to define a uniform harvest area since the bush size varied between the three locations. One measurement was carried out at a height of 1.5 m per bush for seven randomly selected bushes in each treatment (White Shields and War Coulee plots) and on seven seedling bushes (Elbowoods plot) in each treatment in years one and two. In year three, measurements were carried out on five bushes in each treatment.

Analysis of the Juneberry treatment yield data was conducted across cultivar, location, and year. The data were visualized using ggplot2 library in R (ggplot2) using box plots with the median across the box and the whiskers as lines extending from quartile1 and quartile3 to end points that are typically defined as the most extreme data points within Q1 − 1.5 × IQR and Q3 + 1.5 × IQR, respectively. Each outlier outside the whiskers is represented by an individual mark [18]. Dots outside the first and third quartile in figures are values that fall within the expected range of values are retained as a typical observation in agronomic experimentation. An analysis of variance (ANOVA) was performed for Juneberry yield data to study treatments (irrigation and fertilization), and its interactions with cultivar and the location and year. Interaction effects were further analyzed separately to compare each component using the Tukey adjusted mean separation method. The data were analyzed using the 2020 R statistical software packages (R version 4.2) [8].

3. Results

There was no precipitation during the fall and early spring in the three-year period (Table 1). Precipitation during the June flowering and fruiting period was less than 5 cm with the exception of June 2020.

Table 1. Monthly precipitation for McClean County (cm). Data from North Dakota Agricultural Weather Network.

Year	Jan.	Feb.	Mar.	Apr.	May	June	July	Aug.	Sept.	Oct.	Nov.	Dec.	Annual
2019	0.0	0.0	0.0	2.8	4.4	3.7	8.2	7.6	21.3	2.6	0.0	0.0	50.7
2020	0.0	0.0	0.0	0.2	2.4	14.4	2.9	2.8	0.6	0.4	0.0	0.0	23.6
2021	0.0	0.0	0.0	0.5	2.0	3.7	8.2	5.1	1.0	8.0	0.0	0.0	28.6

Figure 1 illustrates the variability in yield between the White Shield and War Coulee location and suggests the presence of cumulative treatment differences. In the 16-year-old cultivar plantings the yield main effects of treatment, location, and year were significant, but the cultivar effect was not (Table 2, Figure 2a). The yield was greater in year two and three than year one for all three treatments across all cultivars and locations, as shown in

Figure 2b. The fertilized treatment showed that the yield increased in year three in contrast to the irrigated treatment (Figure 2b). The interaction effects of treatment by cultivar, treatment by location, and treatment by year were significant (Table 2). At White Shield, the fertilized treatment increased yield in Honeywood and the irrigated and fertilized treatments increased yield in Smokey; however, these did not show a significant increase yield for Martin. At War Coulee, Martin and Smokey showed increased yield in irrigated and fertilized treatments in years two and three.

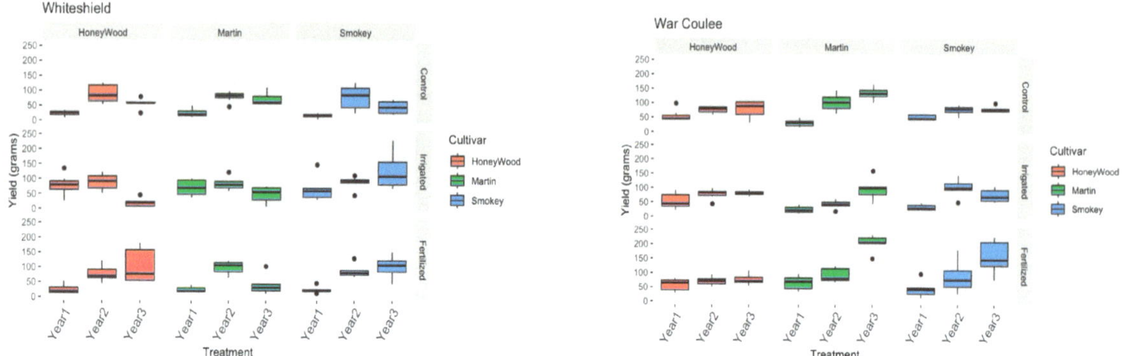

Figure 1. Visualization of treatment yield (g/0.03 m^{-3}) for three cultivars and three years in the 16-year-old Juneberry plantings at White Shield and War Coulee locations. Years 1, 2 and 3 are 2019, 2020, and 2021, respectively. Yield was measured for a constant bush volume (g/0.03 m^{-3}) in seven randomly selected bushes for each treatment in years one and two and five bushes in year three. Boxplots were constructed as described in the Methods section.

Table 2. Yield analysis of variance for main effects (cultivar, treatment, location and yield) and treatment by location, cultivar, and year interactions.

Factor	df	Mean sq	F-Value	*p*-Value
Location	1	9300	7.93	0.005
Cultivar	2	1561	1.33	0.266
Treatment	2	4162	3.55	0.030
Year	2	77,301	65.90	<0.001
Treatment–Location	2	13,884	11.84	<0.001
Treatment–Cultivar	4	4277	3.65	0.006
Treatment–Year	4	6563	5.59	<0.001
Residuals	322	1173		

The Elbowoods seedling plot was three years old at the beginning of this study and had limited fruit in year one (2019), (Figure 3). In 2020, the seedling had yields (g/0.03 m^{-3}) that were approaching those of the mature cultivar plantings and appeared to benefit from irrigation and fertilization (Figure 3). Conditions going into year three were very dry with only 12.9 cm of natural precipitation in the 12 months from July 2020 (harvest) to the end of June 2021. Under these conditions, birds were eating fruit before it was fully ripe and no fruit was available for harvest.

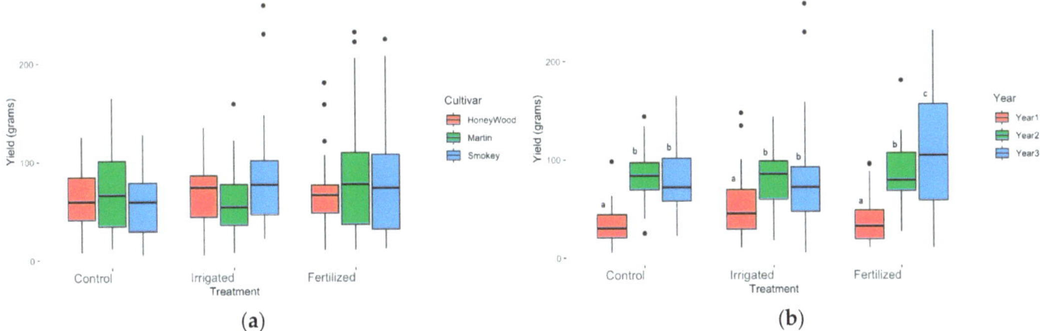

(a)   (b)

Figure 2. Treatment yield ($g/0.03\ m^{-3}$) for cultivars and years across locations. (**a**) Treatment yield across locations and years; (**b**) treatment yield across cultivars and locations. Yield was measured for a constant bush volume ($g/0.03\ m^{-3}$) in seven randomly selected bushes for each treatment in year one and two and five bushes in year three. Boxplots were constructed as described in the Methods section. Significant differences within treatments and among years are noted using different letters. Treatment by year interaction *p*-value < 0.001; n = 21.

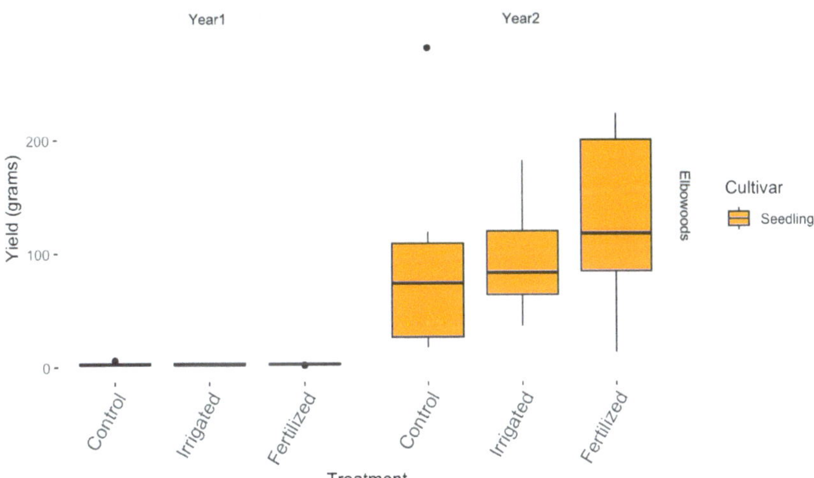

Figure 3. Elbowoods Turner State Nursery Juneberry seedling yields. The plants were three years old in year one (2019). Yield was measured for a constant bush volume ($g/0.03\ m^{-3}$) in seven randomly selected bushes for each treatment in year one and two. Boxplots were constructed as described in methods. Year one was the first year of fruiting and yield was <5 $g/0.03\ m^{-3}$ (n = 7). Treatment effect was not significant in year two (n = 7), and in year three (2021) complete fruit predation eliminated data collection.

4. Discussion

Precipitation accumulation is limited between fall and spring in McClean County, on the Fort Berthold Reservation where the Juneberry cultivar trials were conducted. This restricted the soil water available for fruit development and the flower initiation for the next growing season in this perennial plant. Previous studies on the establishment of Juneberry in North Dakota have shown the importance of supplemental water to prevent young plant loss during the establishment of Juneberry plants [1,19]. Once the plants are established it is suggested that natural rainfall is adequate for the Juneberry maintenance [1,19]. The bushes at White Shield and War Coulee were the same age, had differing soil pH, but similar chemistry and the War Coulee upland and wind exposed site had smaller bushes. The War Coulee was exposed to more wind than the down slope White Shield site. The Elbowoods plants were just coming into production and were only 1.0 m tall. Therefore, all yields are presented for a uniform bush volume (0.03 m^3) rather than a total bush harvest as previously reported for non-irrigated Juneberry production [19]. It is noted that even though a consistent bush volume was used, the location effect was significant and differences in bush structure (i.e., canopy density) could be a contributing factor.

A previously published non-irrigated Juneberry cultivar evaluation indicates that yield varies widely between years and suggests that either precipitation or alternate year bearing (tendency to fruit heavily every other year) could promote this oscillation in yield [19]. There are many studies on Juneberry fruit chemical makeup but few studies on fruit yield in low rainfall conditions. A Canadian ecological study of fruit production in native stands of Juneberry and other native berry shrubs attempted to correlate rainfall and evapotranspiration in the summer with the following growing season's fruit production. A drought code was developed using rainfall and evapotranspiration during the fruiting period to provide an estimate of soil moisture availability, and correlation of the drought code with fruit production in the next growing season was explored [20]. A negative correlation was found between the summer drought code and the following season's production, suggesting the importance of the previous season's precipitation on fruit production in the following season. Although the native Canadian Juneberry fruit production was affected by dry summer weather, it is noted that the prior season fruit production level accounted for more of the annual production variation than the precipitation did. Thus, the previous season's production level and its impact on resource availability contributes to alternate year bearing patterns [20]. Juneberry inflorescence initiation occurs in the axillary buds towards the end of fruit ripening period. Resource competition (such as that related to water or nutrients) during fruiting period completion in the current season could reduce floral development for the next year, as suggested by previous studies [21]. Although precipitation was greater from July to September in 2019 (year 1) than in 2020 and 2021, the yield of the control treatment did not vary significantly across the years. Similarly, comparison of the control and irrigation treatment compared across locations and cultivars in years two and three indicated no significant difference in yield between the control and irrigated treatments. However, the fertilized treatment showed an increased yield with cumulative fertilization in the third year, suggesting that nutrient status is more limiting than the natural precipitation levels.

Despite being impacted by predation in the third year of data collection, a similar response appeared to be developing in the seedling plots. The native seedling materials were developed as an economically feasible alternative to the use of clonally propagated cultivars. Compared to the cost of clonally propagated cultivars, which comes to approximately USD 5/plant plus shipping, seedling plants can be purchased from the state nursery for a more affordable price of USD 1.5/plant. Although the plants are more variable in bush and berry size, the seedling plants from wild gathered berries may be more suitable to a given region. In addition, wild biotype fruits gathered in the same area of cultivar trials are richer in phenolic compounds and antioxidant activity than the fruit from cultivars. This suggests that the use of seedling material may provide a more functional food [8].

Currently there are no standardized recommendations for Juneberry fertilization. In this study, the fertilizer was delivered through the irrigation water to ensure incorporation into the soil since, because early spring precipitation is limited, a spring broadcast of granular fertilizer would not be easily incorporated into the soil. This study indicates that fertilization during the flowering and fruiting periods of 2019 to 2021 resulted in an increased yield in 2021. Future studies would be best served by using new seedling plantings, tracking fruit size and adding additional fertilization treatments, leaf foliar analysis and a fertilization-only treatment. Future studies should also consider pruning strategies to maintain an open sunlight canopy; as nutrient availability increases, it may result in shading within the bush interior and ultimately limit the productive area of the bush. This study suggested that the plants' nutrient availability may be a more limiting factor than that of water alone on the native *Amelanchier alnifolia*. Future studies should focus on exploring timing and amount of water and fertilization needed to develop recommendations for sustainable production. This will be especially important in arid regions where precipitation is limited and trucking water to plantings increases production costs.

Supplementary Materials: The following supporting information can be downloaded at: https://www.mdpi.com/article/10.3390/horticulturae9060653/s1.

Author Contributions: Conceptualization and methodology, K.H. and A.F.; formal analysis, D.A.; writing—original draft, A.F.; writing—review and editing, K.H., D.A. and A.F.; project administration, K.H.; funding acquisition, K.H. and A.F. All authors have read and agreed to the published version of the manuscript.

Funding: This research was funded by the U.S. National Institute of Food and Agriculture, grant number 2018-08417.

Institutional Review Board Statement: Not applicable.

Informed Consent Statement: Not applicable.

Data Availability Statement: Supplemental file.

Conflicts of Interest: The authors declare no conflict of interest.

References

1. Hartman, K.E. Reestablishing the Juneberry on the Fort Berthold Indian Reservation: Cultural, Horticultural, and Educational Connections. Ph.D. Thesis, South Dakota State University, Brookings, SD, USA, 2008.
2. LaDuke, W.; Alexander, S. Food is Medicine: Recovering traditional foods to heal the people. *J. Am. Indian High. Educ.* **2004**, *17*, 17–21.
3. Moerman, D.E. *Native American Ethnobotany*; Timber Press, Inc.: Portland, OR, USA, 1998; pp. 445–448.
4. St-Pierre, R.G.; Zatylny, A.M.; Tulloch, H.P. Evaluation of growth and fruit production characteristics of 15 Saskatoon (*Amelanchier alnifolia* Nutt.) cultivars at maturity. *Can. J. Plant Sci.* **2005**, *85*, 929–932. [CrossRef]
5. Mazza, G. Compositional and functional properties of Saskatoon berry and blueberry. *Int. J. Fruit Sci.* **2005**, *5*, 101–120. [CrossRef]
6. Bakowska-Barczak, A.M.; Kolodziejczyk, P. Evaluation of Saskatoon berry (*Amelanchier alnifolia* Nutt.) cultivars for their polyphenol content, antioxidant properties, and storage stability. *J. Agric. Food Chem.* **2008**, *56*, 9933–9940. [CrossRef] [PubMed]
7. Kraft, T.F.B.; Dey, M.; Rogers, R.B.; Ribnicky, D.M.; Gipp, D.M.; Cefalu, W.T.; Raskin, I.; Lila, M.A. Phytochemical composition and metabolic performance-enhancing activity of dietary berries traditionally used by Native North Americans. *J. Agric. Food Chem.* **2008**, *56*, 654–660. [CrossRef] [PubMed]
8. Poelaert, B.; Bergh, F.; Harman, K.; Reese, R.N. Influence of environment and cultivar on fruit quality in newly established Juneberry (*Amelanchier alnifolia* (NUTT.) NUTT. EX M. ROEM.) orchards on the Fort Berthold Reservation. *Proc. S. Dak. Acad. Sci.* **2010**, *89*, 95–104.
9. Juríková, T.; Balla, S.; Sochor, J.; Pohanka, M.; Micek, J.; Baron, M. Flavonoid profile of Saskatoon berries (*Amelanchier alnifolia* Nutt.) and their health promoting effects. *Molecules* **2013**, *18*, 12571–12586. [CrossRef] [PubMed]
10. Ochmian, I.; Kubus, M.; Dobrowolska, A. Description of plants and assessment of chemical properties of three species from the *Amelanchier* genus. *Dendrobiology* **2013**, *70*, 59–64. [CrossRef]
11. Gunawardana, S.; Eberhart, K.; Hartman, K.; Halaweish, F.T. Assessment of chemopreventive contents of Native American Juneberries (*Amelanchier alnifolia* Nutt. Ex M. Roem.). *Pharm. Crops* **2016**, *6*, 13–21. [CrossRef]
12. Donno, D.; Cerutti, A.K.; Mellano, M.G.; Prgomet, Z.; Beccaro, G.L. Serviceberry, a berry fruit with growing interest of industry: Physicochemical and quali-quantitative health-related compound characterization. *J. Funct. Foods* **2016**, *26*, 157–166. [CrossRef] [PubMed]
13. Hu, C.; Kwok, B.H.L.; Kitts, D.D. Saskatoon berries (*Amelanchier alnifolia* Nutt.) scavenge free radicals and inhibit intracellular oxidation. *Food Res. Int.* **2005**, *38*, 1079–1085. [CrossRef]
14. Wang, J.; Mazza, G. Inhibitory effects of anthocyanins and other phenolic compounds on nitric oxide production in LPS/IFN-gamma-activated RAW 264.7 macrophages. *J. Agric. Food Chem.* **2002**, *50*, 850–857. [CrossRef] [PubMed]
15. Męczarska, K.; Cyboran-Mikołajczyk, S.; Włoch, A.; Bonarska-Kujawa, D.; Oszmiański, J.; Kleszczynska, H. Polyphenol content and bioactivity of Saskatoon (*Amelanchier alnifolia* Nutt.) leaves and berries. *Acta Pol. Pharm.* **2017**, *74*, 660–669. [PubMed]
16. Lanchowicz, S.; Wisniewski, R.; Ochmian, I.; Drzymala, K.; Pluta, S. Anti-microbiological, anti-hyperglycemic and anti-obesity potency of natural antioxidants in fruit fractions of Saskatoon berry. *Antioxidants* **2019**, *8*, 397. [CrossRef] [PubMed]

17. Du Preez, R.; Wanyonyi, S.; Mouatt, P.; Panchal, S.K.; Brown, L. Saskatoon Berry *Amelanchier alnifolia* Regulates Glucose Metabolism and Improves Cardiovascular and Liver Signs of Diet-Induced Metabolic Syndrome in Rats. *Nutrients* **2020**, *12*, 931. [CrossRef] [PubMed]
18. R Core Team. *A Language and Environment for Statistical Computing*; R Foundation for Statistical Computing: Vienna, Austria, 2020; Available online: https://www.R-project.org (accessed on 20 October 2022).
19. Ardayfio, N.K.; Hatterman-Valenti, H. Juneberry cultivar evaluation in North Dakota. *HortTechnology* **2015**, *25*, 747–751. [CrossRef]
20. Howe, E.J.; Obbard, M.E.; Bowman, J. Prior reproduction and weather affect berry crops in central Ontario, Canada. *Popul. Ecol.* **2011**, *54*, 347–356. [CrossRef]
21. Steeves, M.W.; Steeves, T.A. Inflorescence development in *Amelanchier alnifolia*. *Can. J. Bot.* **1990**, *8*, 1680–1688. [CrossRef]

horticulturae

MDPI

Article

Effects of Different Irrigation Rates on Remontant Strawberry Cultivars Grown in Soil

Micol Marcellini [1], Davide Raffaelli [1], Luca Mazzoni [1], Valeria Pergolotti [1], Francesca Balducci [1], Yasmany Armas Diaz [2], Bruno Mezzetti [1,3] and Franco Capocasa [1,*]

1 Department of Agricultural, Food and Environmental Sciences, Università Politecnica delle Marche (UNIVPM), Via Brecce Bianche 10, 60131 Ancona, Italy
2 Department of Clinical Sciences, Polytechnic University of Marche, 60131 Ancona, Italy
3 UNEA, Research Group on Food, Nutritional Biochemistry and Health, Universidad Europea del Atlántico, Isabel Torres, 21, 39011 Santander, Spain
* Correspondence: f.capocasa@staff.univpm.it; Tel.: +39-0712204640

Abstract: The present study assessed the responses, in terms of vegetative, productive, qualitative, and nutritional features, of plants and berries of three remontant strawberry cultivars cultivated in soil and irrigated using three irrigation regimes: standard irrigation regime (W100), 20% (W80) less irrigation than the standard irrigation, and 40% (W60) less irrigation than the standard irrigation. The tested plants were "Albion", "San Andreas", and "Monterey", which were cultivated in the east coast area of Marche, Italy. Specifically, the study examined the response of the genotype to irrigation deficit, highlighting the performance of the "Monterey" cultivar, which showed improvement in terms of fruit firmness, folate content, and antioxidant capacity at the W80 irrigation regime without a significant yield reduction. In all the cultivars, when irrigation was reduced by up to 20% of the standard irrigation regime (W100), there were no significant losses of yield or reduction in the fruits' sensorial quality or antioxidant activity. The results showed that the standard irrigation regime (W100) commonly adopted by the farmers in the Marche area uses more water than necessary. With more accurate water management, it will be possible to save almost 226 m^3 of water per hectare per cultivation cycle.

Keywords: strawberry; water stress; remontant; sensorial quality; nutritional compounds; soil

Citation: Marcellini, M.; Raffaelli, D.; Mazzoni, L.; Pergolotti, V.; Balducci, F.; Armas Diaz, Y.; Mezzetti, B.; Capocasa, F. Effects of Different Irrigation Rates on Remontant Strawberry Cultivars Grown in Soil. *Horticulturae* **2023**, *9*, 1026. https://doi.org/10.3390/horticulturae9091026

Academic Editor: Toktam Taghavi

Received: 11 July 2023
Revised: 30 August 2023
Accepted: 5 September 2023
Published: 11 September 2023

1. Introduction

In recent decades, the strawberry (*Fragaria × ananassa* Duch.) has garnered much interest because of its various features [1]. Consumers appreciate this fruit because of the color, shape, taste, and nutritional properties of the berry [2–9]. However, the current changes in the climate, leading to extreme weather conditions, for example, drought [10], have increased the stress on the plant, leading to uncertainty in the strawberry market. According to the FAO 2021 [11], the annual worldwide water withdrawal from natural water bodies was about 4250 km^3, where agriculture used 71.7% of the total water consumed. By 2050, the human population is estimated to reach nine billion; to feed this population, agriculture will have to cope with several challenges [12]. About 60% extra food will be needed, implying an even higher water consumption [13]. Recently, the Sustainable Development Goal (SDG) 6.4.2 [11] was created to evaluate a country's water stress level. It considers the ratio of the total freshwater withdrawn by all major sectors to the difference between the total renewable freshwater resources and the environmental water requirements, multiplied by 100 [14]. Between 2015 and 2018, Italy achieved 30% of S.D.G. 6.4.2, demonstrating a low level of water stress [15]. The S.D.G. of 2022 is not available yet. Nevertheless, in the January–June semester, with +0.76 °C, Italy faced the warmest period ever recorded [16]. In this scenario, in agriculture, the level of water consumption must be maintained within certain limits, avoiding water abuse and groundwater contamination [17,18]. Increased temperature negatively influences the plants' growth and development [19,20], leading to stunted reproductive organs because of lowered carbon assimilation. Many studies have highlighted the importance of selecting the optimal performing genotype to save water in strawberry cultivation [21]. New breeding programs should select new cultivars with reduced water need [22]. Furthermore, appropriate studies on cultivar, environment, and cultivation system interactions should be developed to define the proper irrigation regimes for more sustainable cultivation protocols and better fruit quality. Regarding strawberries, particular attention is now being paid to the development of cultivation systems that can promote out-of-season production using remontant cultivars able to fruit in different climatic conditions without the need for the winter season and in different growing conditions. Therefore, there is a need to identify appropriate cultivation conditions for remontant strawberry cultivars. In this work, we tested the response of three strawberry remontant cultivars, grown in open-field conditions and adopting reduced irrigation regimes, with the aim to develop the most sustainable and quality production practice.

2. Materials and Methods

2.1. Plant Materials

This one-year-cycle experiment was set in the experimental farm of the regional extension service (Agenzia Servizi al Settore Agroalimentare delle Marche, ASSAM) located in Petritoli, Italy (43°03′10″, 13°41′20″). We used remontant strawberry cultivars grown in open fields for a single production. On 24 April 2019, frigo plants were planted in soil and covered by a plastic tunnel and the fruits were collected in the summer of 2019. The studied cultivars were frigo plants "Albion" (A+), "San Andreas" (A++), and "Monterey" (A+), three remontant cultivars well known for their consistent productivity during the season ("Albion"); earliness, rusticity, and quality fruit ("San Andreas"); and yield, quality, and resistance to diseases ("Monterey") [23].

2.2. Experimental Design and Irrigation Scheduling

The plants were planted in double rows. The plants in each row were 30 cm apart and the rows were 35 cm apart, resulting in a density of 5.5 plants m^{-2}. The plants were grown in non-fumigated, chalky, and high-pH soil, as described in Table 1. The fertigation program, controlled by a Dosatron® D8R (Dosatron SAS, Tresses, FR), involved the distribution of N (120 unit ha^{-1}), P (100 unit ha^{-1}), and K (150 unit ha^{-1}) during the cultivation cycle with daily treatment. Each line had two dripline hoses Toro® Acqua-Traxx with a 1.1 L hour^{-1} flow rate. For the cultivation, we followed the standard integrated pest management (IPM) (Directive128/2009). Before the start of the irrigation treatment, all plants received the same amount of water (1378 m^3 ha^{-1}) to ensure good establishment of the plant. We started the experimental irrigation at the flowering stage (stage 6 BBCH) and ended it on the last harvest date (stage 8 BBCH). Three irrigation treatments (W) were applied: W100 (control) with an irrigation rate suggested by the Marche Region Directive 786 on 10 July 2017 [24], corresponding to 1183 m^3 ha^{-1}, and W80 and W60, with 20% and 40% less water used, respectively, corresponding to 957 m^3 ha^{-1} and 665 m^3 ha^{-1} of total water used for irrigation by the end of the experiment. The soil humidity was monitored by six tensiometers, two per treatment, and placed at a 15 cm depth, approximately the root exploration area. The moisture probes, Watermark®, were characterized by a datalogger that took daily measurements (Figure S1). The temperature over the experimental period was monitored through the ASSAM weather station (Figure S2). The split-plot design of the experimental field consisted of three main blocks, differentiated by three different water supply levels, repeated for "San Andreas", "Albion", and "Monterey" cultivars. Each cultivar represented a sub-block and was composed of three replicates, called "plots", consisting of 8 plants each, for a total of 27 plots and 216 plants (3 blocks × 3 cultivars × 3 replicates) as shown in Figure S3.

Table 1. Soil feature of the Agenzia Servizi Settore Agroalimentare delle Marche (ASSAM) experimental field.

Soil Parameter	Unit	Results	Method
pH		8.14	[25]
Sand	$g\,Kg^{-1}$	304	[26]
Silt	$g\,Kg^{-1}$	399	[26]
Clay	$g\,Kg^{-1}$	297	[26]
Active limestone	$g\,Kg^{-1}$	61	[26]
Total limestone	$g\,Kg^{-1}$	174	[26]
Assimilable P	$g\,Kg^{-1}$	3.7	[27]
Exchangeable Na	$g\,Kg^{-1}$	15	D.M. 13/09/99 GU SO n.248 del 21/10/1999 III.2, XIII.2.6
Cation exchange capacity	$mEQ\,100\,g^{-1}$	21.9	D.M. 13/09/99 GU SO n.248 del 21/10/1999 III.2
Assimilable iron	$g\,Kg^{-1}$	9.7	[28]
Assimilable M n	$g\,Kg^{-1}$	4.1	[29]
Assimilable Z n	$g\,Kg^{-1}$	0.52	[29]
Assimilable C u	$g\,Kg^{-1}$	2.7	[29]
Boron soluble		0.1	[30]
C/N		7.7	
Organic matter	$g\,Kg^{-1}$	11.9	D.M. 13/09/99 GU SO n.248 del 21/10/1999-VII.3. VII.3.6
Total N	$g\,Kg^{-1}$	0.90	[31]
Mg/K		2.7	
Exchangeable Mn	$mg\,Kg^{-1}$	155	[29]
Exchangeable K	$mg\,Kg^{-1}$	410	[32]

2.3. Plant Growth and Vegetative Parameters

The leaf number and plant height were recorded three times during the season. The measurements were taken on 7 July 2019, 7 August 2019, and 7 September 2019. One measurement date (7 August 2019) was applied for the number of crowns, inflorescences number, leaf length, and leaf width.

2.4. Fruit Production

Strawberries were harvested on 11 dates: 2 July, 9 July, 15 July, 22 July, 29 July, 5 August, 12 August, 19 August, 26 August, 2 September, and 9 September. To evaluate the ripening stage, we used the methods described by Capocasa et al. [33] and a precocity index (IP), which represents the average number of weighted days needed to collect the whole production of a cultivar from 1 January. The other parameters were the average fruit weight (AFW), the total yield, and the marketable production (fruits $\geq$ 22 mm and not rotted or deformed).

2.5. Fruit Quality

For each harvest date, 10 fully ripe strawberry fruits were collected from each plot. Fruits for the qualitative analyses, both organoleptic and nutritional, were selected from the first, second, and third main pickings. We collected the fruits from six plants at the center of each plot and pooled together the fruits deriving from the three replicates of each cultivar. The collected strawberries were fully ripe, without any visible injuries, and of a homogenous size.

2.5.1. Fruit Organoleptic Quality

Ripe fruits were analyzed for color, firmness, total soluble solids, and titratable acidity in accordance with Marcellini et al. [34]. For each thesis (genotype/treatment), at each harvest, we selected 10 fruits to evaluate the chroma, also known as color saturation (Minolta Chromameter CR 400, Konica Minolta, Tokyo, Japan) and firmness (Penetrometer 327, Effegì, Ravenna, Italy). To evaluate the external color of fresh fruits, the CR-400 was used, measuring two points on opposite sides of each fruit using CIELAB values (L*, a*, b). The chroma was evaluated from a and b values. The genotype and the ripeness stage influenced the chroma value. Next, we perforated the same fruits using the penetrometer, through a 6 mm star probe. Until the total soluble solids (TSS) and titratable acidity (TA) evaluation, the samples were frozen at -18 °C. A soluble solids measurement was performed using a digital refractometer (PR-101α ATAGO, Tokyo, Japan) for TSS and acid–base titration was carried out for TA. The TA was calculated as mEQ of NaOH per 100 g of fresh weight (FW) as follows: on 10 g of strawberry juice as the base, we added 10 g of distilled water and a few droplets of bromothymol blue (pH indicator) with a 0.1 N NaOH solution. The final acidity content was expressed as described by the following formula [35]:

$$\% \text{ acid} \left(\frac{\text{wt}}{\text{wt}} \right) = \frac{\text{N} \times \text{V} \times \text{Eq. wt.}}{\text{W} \times 1000} \times 100$$

where

N = normality of the titrant, NaOH (mEQ/mL)
V = volume of the titrant (mL)
Eq. wt. = equivalent weight of the predominant acid (mg/mEQ)
W = mass of the sample (g)
1000 = factor relating milligrams to grams (mg/g) (1/10 = 100/1000).

2.5.2. Fruit Nutritional Quality

We stored the strawberries and harvested for the analysis of nutritional compounds in plastic bags at -18 °C in laboratory freezers until the day of the extraction. For the extraction, we followed the method described by Mezzetti et al. [36]. In short, from each bag, we chose five strawberries and cut each fruit into four pieces: for the analysis, we used only half of the fruit, the part derived from opposite faces of the fruit, so as to avoid any bias connected to the influence of sunlight during cultivation. The strawberry pieces were chopped and weighed: 10 g was designated for the methanolic extract suitable for detecting phenolic acids, polyphenols, anthocyanins, and antioxidant capacity; 1 g for extracting vitamin C; and 2 g for extracting folates. After the extraction, the fruit samples were analyzed by two methods: HPLC, to detect ascorbic acid, folates, and phenolic acids, as well as spectrophotometry, to evaluate polyphenols, anthocyanins, and antioxidant capacity.

2.5.3. HPLC

Both ascorbic acid content and folate content were quantified as described by Mezzetti et al. [28]. For vitamin C, we added 4 mL of the extraction buffer made of Milli-Q water, 5% meta-phosphoric acid, and 1 mM ethylenediaminetetraacetic acid to 1 g of strawberry sample, homogenized the mixture using an Ultraturrax T25 homogenizer (Janke and Kunkel, IKA Labortechnik, Staufen, Denmark), and sonicated the same for 5 min. After centrifuging the samples (2500 rpm at 4 °C, for 10 min), we filtered them through a 0.22 μm nylon syringe filter. The samples with the vitamin C extracted were analyzed in the HPLC system, specifically, a Jasco PU-2089 Plus controller (Jasco Inc., Easton, MD, USA) and a Jasco UV-2070 Plus ultraviolet (UV) (Jasco Inc., Easton, MD, USA) detector set at an absorbance of 260 nm. The HPLC column used was an Ascentis Express C18 150 × 4.6 mm (Supelco, Bellefonte, PA, USA), protected by a Phenomenex 4.0 × 3.0 mm C18 ODS guard column (Phenomenex, Torrance, CA, USA). The calibration curve was prepared by the standard concentration of the vitamin. Finally, the unit of measurement was mg Vit C per 100 g of fruit weight (FW) from three replications per sample.

For folate extraction, we added 2 g of frozen strawberries to 8 mL of the extraction buffer (0.1 M phosphate buffer containing 1.0% of L(+)-ascorbic acid (w/v) and 0.1% 2,3-dimercapto-1-propanol (v/v) at pH 6.5, freshly prepared) and homogenized the mixture using an Ultraturrax T25 homogenizer (Janke and Kunkel, IKA Labortechnik, Staufen, Denmark) at a high speed. The falcon tube containing the sample was immersed in a water bath at 100 °C for 10 min and then rapidly cooled at −18 °C. Next, to deconjugate polyglutamylation folates, we added 150 µL of folate conjugase from the hog kidney to each sample and incubated the mixture in a shaking oven at 37 °C for 3 h. We placed the tube in a thermal bath at 100 °C for 5 min and then rapidly cooled the mixture at −18 °C and centrifuged it at 4000 rpm at 4 °C for 20 min. The supernatant was collected in a 25 mL falcon tube. Then, the pellet was reprocessed with the same extraction buffer, warmed in a water bath, cooled, and centrifuged. We added the supernatant to the extracted sample and filled the falcon tube with the extraction buffer so that the volume became 25 mL. We filtered the samples (0.45 µm syringe filter) as described by Iniesta et al. [37] and Jastrebova et al. [38] with some modifications. The filtrates were purified through solid-phase extraction on anion-exchange Isolute cartridges. Finally, we carried out the HPLC analysis in accordance with Strålsjö et al. [39] with some modifications. The analytical column was a Luna C18, 250 × 4.6 mm, 5 µm (Phenomenex, Torrance, CA, USA), protected by a Phenomenex 4.0 × 3.0 mm C18 ODS guard column (Phenomenex, Torrance, CA, USA). A fluorescence detector (FLD) FP-2020 Plus (Jasco, Easton, MD, USA) set at wavelengths of 290 nm excitation and 360 nm emission and an autosampler AS-4050 (Jasco, Easton, MD, USA) were also used. The results were expressed as µg of folate per 100 g of FW. The results were obtained with three replications ± standard deviation.

For phenolic acid analysis, the procedure adopted was in accordance with Frederick et al. [40]. The HPLC setup consisted of a Jasco PU-2089 plus controller, a Jasco UV-2070 plus ultraviolet detector, and a Jasco AS-4050 autosampler, all from Jasco (Easton, MD, USA). The chromatographic column employed was an Aqua Luna C18, with dimensions of 250 mm × 4.6 mm, manufactured by Phenomenex (Torrance, CA, USA). This column was safeguarded by a Phenomenex 4.0 mm × 3.0 mm C18 ODS guard column. The separation process involved a gradient program with two mobile phases: A (containing 2% acetic acid) and B (composed of acetic acid, acetonitrile, and water in a ratio of 1:50:49). The gradient initiated with 55% A and 45% B for 50 min, followed by a 10-min phase of 100% B. Subsequently, it was reduced to 10% B until the analysis concluded. To quantify and recognize only phenolic acids, the UV/VIS detector was set to 320 nm. Three standard solutions were prepared from the following pure phenolic acids: chlorogenic acid, caffeic acid, and ellagic acid. For caffeic and chlorogenic acids, ethanol (C_2H_6O) was used as a solvent. For ellagic acid, sodium hydroxide (NaOH, 1 M) was used as a solvent. The results were expressed as mg of phenolic acids per 100 g of fresh fruit.

2.5.4. Spectrophotometry

We measured the anthocyanin content (ACY) using the pH differential shift method [41]. Each sample was diluted at a ratio of 1:10 with two solutions: potassium chloride (pH 1.00) and sodium acetate (pH 4.50). Then, the absorbance for both solutions was measured at 500 and 700 nm. The data were expressed as mg pelargonidin-3-glucoside (molar extinction coefficient 15,600 L mol^{-1} cm^{-1}; molecular weight 433.2 g mol^{-1}) per kg of FW. The total antioxidant capacity (TAC) was measured through the 2,2′-azino-bis (3-ethylbenzothiazoline-6-sulphonic acid) (ABTS) assay [42,43]. ABTS is a colorless substance that turns into the colored monocationic radical form when exposed to an oxidative agent. The extent of decolorization of the monocationic radical form is a function of the antioxidants present in the strawberries and was calculated relative to the reactivity of Trolox, a water-soluble vitamin E analog. Antioxidant activity is expressed as mg Trolox equivalent per kg of FW and the results were expressed as the mean of six replications ± standard deviation. The total polyphenol content (TPH) was measured using the Folin–Ciocalteu reagent method [44]. Briefly, we filled a glass test tube with 7.0 mL of water. To this, we first added 1 mL of the diluted sample (1:20) and then added 500 mL of the 2 N Folin–Ciocalteu reagent (Sigma-Aldrich, St. Louis, MO, USA). The solution was vortexed and allowed to react for 3 min. Then, we added 1.5 mL of a 20% sodium carbonate solution. The contents of the tube were mixed again and the tube was stored in the dark for 60 min. After this, we measured the sample absorbance at 760 nm. The data were expressed as mg gallic acid per kg of FW. The results were obtained as the mean of six replications ± standard deviation.

2.6. Statistical Analysis

The results are presented as the values ± standard deviation and were subjected to a two-way analysis of variance (ANOVA) at confidence levels of 95% and 99%. Significant differences were calculated according to Fisher's LSD test and differences at $p < 0.05$ were significant. Principal component analysis (PCA) was also used to evaluate the levels of association among the nutritional parameters. Statistical analyses were performed using Statistica 7 software (StatSoft, TIBCO Software, Palo Alto, CA, USA).

3. Results and Discussion

3.1. Vegetative Parameters

Considering the vegetative parameters, in "Albion" and "San Andreas" there was an increase in the branch crown number at high and moderate water supply (W100 and W80, respectively), while in "Monterey", there was a significant decrease in this parameter at a slight water shortage (W80). On severe water reduction (W60), the number of branches per plant was reduced in comparison with what was detected in plants grown at W100, like the results provided by Gehrmann [45], Awang et al. [46], and Marcellini et al. [34], underlining the negative influence of the salinity arising from the water shortage on the vegetative apparatus of the strawberry plant. The number of inflorescences in the plants was quite low, even at full irrigation. In addition, in this case, "Monterey" showed a significant sensitivity to water shortage at W60 compared to W100 (Table 2). As already described [47,48], water shortages tended to influence the vegetative structure of the plants, indicated by the lower plant height and leaf number, possibly leading to a reduction in photosynthesis [49]. Therefore, when the water supply is low, the plant's habit is more compact. For all treatments, "S. Andreas" was the most vigorous, with the highest value in terms of the plant height and the number and size of leaves, followed by "Monterey" and "Albion". "Monterey" had fewer branch crowns and inflorescences at reduced irrigation and reduced plant height at W80. Generally, "Albion" was smaller in height than other cultivars, with a particularly negative influence at W60 (Table 2). The leaf number was also significantly reduced at W60. The different treatments did not influence the leaf size in any of the cultivars (Table 2). The development of the plants in terms of plant height and leaf number was monitored between July and September 2019 (Figure 1). We noted that for the three cultivars examined, the leaf number increased, particularly at W100 and W80, while at W60 during the summer season, this parameter remained constant. Contrarily, the plant height tended to decrease slightly for all treatments.

Table 2. Effects of water availability on branch crown number, the number of inflorescences, leaf length, and leaf width in different strawberry cultivars.

	Cultivar		
Number of branch crowns	Albion	Monterey	S. Andreas
W100	2.3 ± 0.8 cd	2.6 ± 0.9 bc	3.2 ± 0.8 a
W80	2.6 ± 1.0 b	2.2 ± 0.8 d	3.1 ± 1.0 a
W60	2.3 ± 0.7 cd	2.3 ± 0.8 cd	2.7 ± 0.7 b
Number of inflorescences			
W100	2.2 ± 2.4 bc	2.7 ± 3.0 a	2.0 ± 3.2 cd
W80	2.5 ± 2.0 ab	2.5 ± 2.6 ab	1.8 ± 2.7 d
W60	2.0 ± 2.7 cd	2.3 ± 2.3 bc	2.0 ± 3.4 cd
Leaf length (cm)			
W100	7.3 ± 1.1 bc	7.0 ± 1.1 cd	7.9 ± 1.2 a
W80	7.3 ± 1 bc	6.8 ± 1.1 d	7.9 ± 1.3 a
W60	7.4 ± 1.0 b	7.0 ± 1.0 cd	7.8 ± 0.9 a
Leaf width (cm)			
W100	6.7 ± 0.8 bc	6.7 ± 1.1 bc	7.2 ± 1.0 a
W80	6.9 ± 1.3 ab	6.5 ± 0.9 c	7.2 ± 0.8 a
W60	6.8 ± 0.8 bc	6.7 ± 1.0 bc	7.0 ± 1.0 ab

Note: Values with the same lowercase letter for the same parameter were not statistically different in Fisher's LSD test ($p < 0.05$). Values are expressed as the means of one year (2019) ± standard deviation.

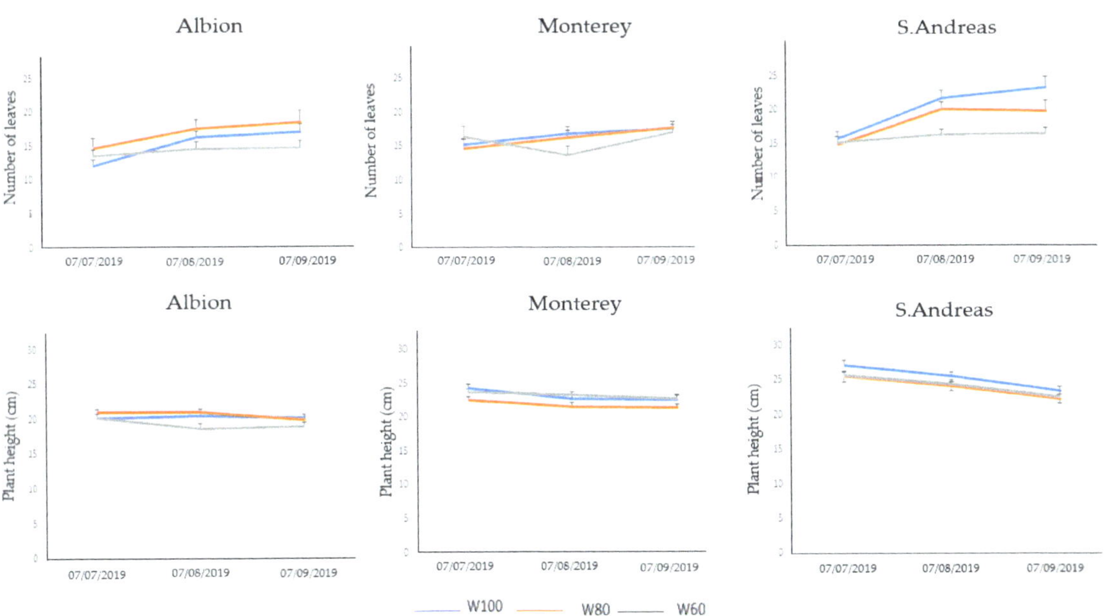

Figure 1. The number of leaves and plant height during water reduction (from July to September 2019).

3.2. Productive Parameters

Generally, reduced water restitution led to the fruits maturing earlier than usual. In fact, at full irrigation (W100), the three cultivars ripened at the same time. However, from

W100 to W60, the fruits matured 6 days earlier for "Albion", 10 days earlier for "Monterey", and 12 days earlier for "San Andreas" (Table 3). On average, the difference in the harvest time was 5 days earlier for W80 and 9 days earlier for W60 with respect to the control treatment (W100). The average fruit weight appeared to be independent of the difference in the amount of water restitution. The W80 treatment negatively influenced the AFW (even when compared to the W60 treatment), though the difference was not statistically significant. Concerning commercial production, "Albion" appeared to be the most sensitive cultivar, with a reduction in 37% in the commercial yield (statistical difference), followed by "San Andreas" (-34%) and "Monterey" (-21%), but without a significant difference (Table 3). For total production too, a similar trend was detected. Compared to W100, in "Albion", the total production was reduced by about 60 g; in "Monterey", the total production was reduced by about 54 g; and in "S. Andreas", the total production was reduced by about 55 g. Among the cultivars, "Monterey" showed the highest commercial and total production for all treatments. As already known from the literature [50–54], proper irrigation treatment is a prerequisite for exploiting the yield potential of a cultivar. In fact, the differences in the production among cultivars in the same water condition demonstrate that efficient water use is genotype-dependent [47,48].

Table 3. Effects of water availability on precocity index and average fruit weight in different strawberry cultivars.

	Cultivar		
Precocity index (days)	Albion	Monterey	S. Andreas
W100	214.2 ± 2.4 ab	214.6 ± 3.4 ab	215.1 ± 6.3 a
W80	209.4 ± 1.4 abcd	209.7 ± 0.1 abc	211.1 ± 4.1 ab
W60	208.6 ± 4.5 bcd	204.4 ± 2.7 cd	203.6 ± 2.8 d
Average fruit weight (g)			
W100	11.0 ± 0.1 cd	12.2 ± 1.8 abc	13.4 ± 0.8 a
W80	10.5 ± 0.5 d	10.5 ± 0.9 d	11.6 ± 1.4 bcd
W60	11.1 ± 0.4 cd	12.1 ± 1.2 abcd	12.9 ± 0.5 ab
Commercial production (g/plant)			
W100	162.9 ± 24.8 ab	181.2 ± 42.3 a	122.1 ± 11.4 bcd
W80	122.6 ± 26.2 bcd	174.0 ± 44.2 a	79.2 ± 26.5 d
W60	103.1 ± 15.9 cd	142.3 ± 17.1 abc	80.9 ± 22.0 d
Total production (g/plant)			
W100	204.8 ± 26.1 ab	241.8 ± 32.4 a	159.4 ± 14.9 bc
W80	164.5 ± 26.0 bc	206.3 ± 50.8 ab	107.5 ± 26.8 d
W60	144.7 ± 31.7 cd	187.9 ± 15.0 bc	104.5 ± 19.2 d

Note: Values with the same lowercase letter for the same parameter were not statistically different in Fisher's LSD test ($p < 0.05$). Values are expressed as the means of one year (2019) ± standard deviation.

3.3. Qualitative Parameters

In fruits, there is a positive relationship between reduced irrigation and the increased content of soluble solids [49–51]. In our experiment, the sugar content of the fruits did not differ from that shown in previous works, highlighting a positive correlation between less water restitution and fruits' °Brix content (Table 4). The cultivars "Monterey" and "Albion" stood out for the high sugar content in their fruits, which increased, respectively, by 1.2 °Brix and 1 °Brix at W60 with respect to W100. In all the treatments, the fruits of "S. Andreas" showed a lower sugar content in comparison with the other cultivars; however, they also showed a significant increase in the content of soluble solids with a lower irrigation supply. At the lower water restitution, the fruits of stressed plants accumulated higher SSC, probably in terms of higher concentrations of fructose and glucose [47]. Regarding acidity, the common tendency among the fruits was the lack of a correlation between the water administrated and the acid content. In each irrigation treatment, the fruits of "Albion" had the highest acidity, followed by those of "Monterey" and "S. Andreas", with similar values. The fruit firmness increased when irrigation was reduced by 20% (Table 4). These results are different from those previously obtained by Krüger et al. [48], where the fruit firmness decreased in a reduced water regime. In the present study, when grown in an optimal condition of irrigation (W100), the cultivars showed an appreciable fruit firmness: "Monterey", 422.1 g; "Albion", 371.0 g; and "S. Andreas", 338.9 g. However, when "S. Andreas" and "Albion" were treated at W60, the fruit firmness increased by 20 g and 25 g, respectively. "Monterey" demonstrated a constancy in terms of fruit firmness. At full irrigation (W100), the chroma value was higher in the fruits of "S. Andreas": they seemed brighter than the other tested cultivars (Table 4). "Albion" and "Monterey" were pretty much the same. The brightness of the fruits of "Albion" and "Monterey" increased in higher-water-stress conditions in contrast to those of "S. Andreas". Our findings were partially in accordance with those of other studies [45,46] but in contrast to the results of the study by Adak et al. [47], where water stress did not affect the fruit color parameters.

Table 4. Effects of water availability on sugar content, titratable acidity, firmness, and chroma in different strawberry cultivars.

	Cultivar		
Sugar content (°Brix)	Albion	Monterey	S. Andreas
W100	14.1 ± 1.2 c	14.5 ± 0.8 bc	11.2 ± 0.7 e
W80	14.6 ± 1.2 bc	15.2 ± 1.1 ab	11.4 ± 0.8 e
W60	15.1 ± 1.3 ab	15.7 ± 1.4 a	12.3 ± 0.8 d
Titratable acidity (mEQ of NaOH/100 g of fruit weight)			
W100	14.9 ± 0.8 a	13.2 ± 1.1 b	13.5 ± 1.3 b
W80	14.5 ± 1.7 a	13.3 ± 1 b	13.2 ± 1.0 b
W60	14.5 ± 1.3 a	13.3 ± 1.2 b	13.5 ± 1.0 b
Firmness (g)			
W100	371.0 ± 94.2 c	422.1 ± 105.7 a	338.9 ± 78.2 d
W80	391.9 ± 88.9 b	419.7 ± 110.3 a	361.4 ± 100.1 c
W60	395.6 ± 91.4 b	420.3 ± 121.5 a	359.1 ± 105.5 c
Chroma			
W100	43.7 ± 5.5 cd	43.2 ± 6.5 d	45.6 ± 5.7 a
W80	44.4 ± 5.0 bc	44.2 ± 5.8 bc	44.8 ± 5.9 ab
W60	45.4 ± 5.3 a	44.4 ± 5.9 bc	44.5 ± 5.9 bc

Note: Values with the same lowercase letter for the same parameter were not statistically different in Fisher's LSD test ($p < 0.05$). Values are expressed as the means of one year (2019) ± standard deviation.

3.4. Nutritional Parameters

The fruit content of ascorbic acid seemed to depend on the interaction between the plant genotype and water restitution to the plants (Table 5). Briefly, reduced water content negatively affected vitamin C accumulation in "Monterey" fruits (26.63 mg/100 g of FW at W100 compared to 22.18 mg/100 g of FW at W60). However, the fruits of "Albion" and "San Andreas" did not show any significant variation between treatment at W100 and at W60, even though they performed worse at W80. On average, "Monterey" performed as the best cultivar in terms of fruit vitamin C content.

Concerning the vitamin B9 fruit content, in all the tested genotypes, reduced water restitution tended to stimulate fruit production with increased folate content (Table 5). "Monterey" plants produced fruits with a significantly higher folate content at W60 thesis than at W100. On average, "Monterey" fruits expressed the highest concentration of folates compared to the other cultivars. "Albion" fruits did not exhibit significant differences among the thesis and "San Andreas" accumulated the highest concentration of vitamins at W60.

In the fruits, the accumulation of total phenolic compounds was not significantly influenced by water regimes or by their interaction with genotype (Table 5), as was also described by Martínez-Ferri et al. [53]. The highest accumulation was detected in fruits from "San Andreas" plants at W60, with 422.50 mg GA/100 g of FW, and the lowest accumulation was detected in fruits harvested from "Albion" plants at W80, with 351.38 mg GA/100 g. Differently from these cultivars, "S. Andreas" plants produced fruits with a lower accumulation of TPH at W100 than at W80 and W60. The same response in terms of fruit anthocyanins content was also detected in plants treated with reduced water regimes, where no significant differences were detected among treatments. Among the cultivars, the fruits of "San Andreas" had the highest amount of fruit anthocyanins (Table 5).

The antioxidant capacity of strawberry fruits harvested from plants treated with reduced water regimes was higher than that of fruits harvested at 100% water restitution (Table 5). Even though this trend was confirmed in each cultivar, only "S. Andreas" presented significant differences, with the fruits harvested at W60 treatment presenting a significantly higher value of TAC (599.23 mg Trolox eq/100 g of FW) than those harvested at W100 (506.66 mg Trolox eq/100 g of FW). The cultivar-dependent effect for this important fruit trait has already been described by Cardeñosa et al. who showed that the "Primoris" cultivar exhibits higher fruit antioxidant capacity under higher saline conditions because of the plant's response to abiotic stress [54].

Concerning the phenolic acid content, different water regimes did not significantly influence the amount of these compounds in strawberry fruits, while, again, the genotype was a determinant factor. In fact, fruits of "Monterey" had the highest content of phenolic acids, presenting the highest amount at the W60 trial. In this case, both "Monterey" and "S. Andreas" displayed a similar trend, showing increasing fruit concentrations of phenolic acids with decreasing water supply. Furthermore, the fruits of "S. Andreas" showed a significant difference in terms of the phenolic acid content detected at W60 (30.52 mg/100 g of FW) and that detected at W100 (27.37 mg/100 g of FW) (Figure 2). Among phenolic acids, strawberry fruits showed the highest concentration of ellagic acid, followed by chlorogenic acid; the low quantity of caffeic acid did not seem to be influenced by reduced water supplies.

To investigate whether one or more nutritional compounds are linked together in their determination, we analyzed the data obtained for the three cultivars using principal component analysis (PCA) (Figure 3). The two main factors reported on the graph justify the 60% of the variability registered in this study. It is interesting to note that the vectors TAC and TPH fall close to each other, indicating a strong relationship between the amount of total phenolics and the antioxidant capacity of fruits. This result confirms the finding that the phenolic compounds are mainly responsible for the TAC of strawberries. Furthermore, the vector phenolic acids is placed in the third quadrant and this was also expected, given the antioxidant capacity exerted by this class of compounds. What is surprising is that the vector folates is in this quadrant, even though these compounds are not strong antioxidants. The vectors ACY and Vit C are placed in opposite quadrants (second and fourth, respectively), indicating that high amounts of one of them in the strawberry fruits of this study corresponded to low amounts of the other and vice versa. However, they are both good antioxidant compounds and they are equidistant from the TAC vector.

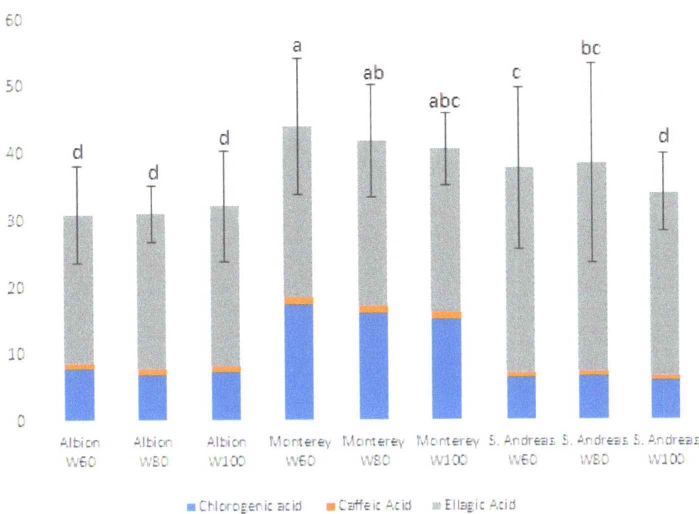

Figure 2. Phenolic acid content (mg 100 g^{-1} of FW) based on different water supplies. Values are expressed as the mean total phenolic content of one year (2019) $\pm$ standard deviation. Different letters indicate significant differences (Fisher's LSD test; $p < 0.05$).

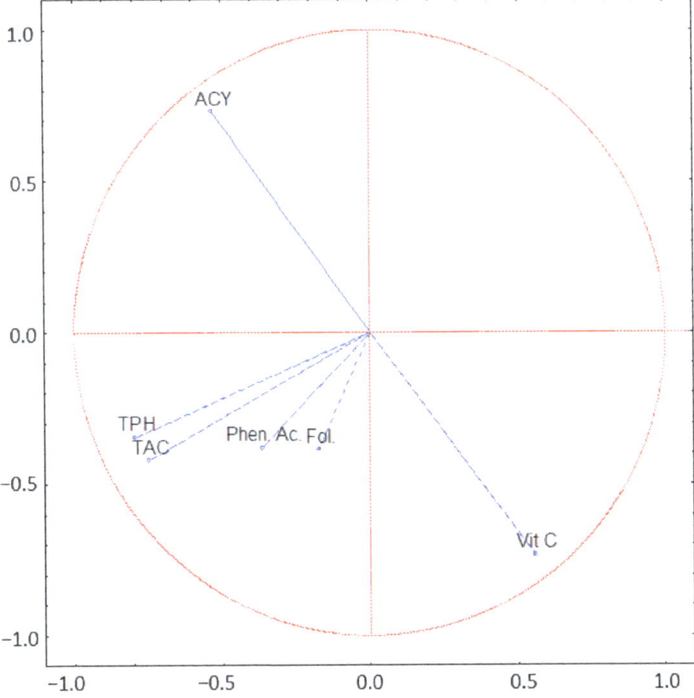

Figure 3. Principal components analysis (PCA). Factor 1: 32.50% and Factor 2: 27.68%. The abbreviations referred to ACY: anthocyanins; TPH: total phenolics; TAC: total antioxidant capacity; Phen.A.: phenolic acids; Fol: folates; Vit C: vitamin C.

Table 5. Effects of water availability on vitamin C, folates, total phenolics, anthocyanins, and total antioxidant capacity concentration in different strawberry cultivars.

		Cultivar	
Vitamin C (mg 100 g^{-1} of FW)	Albion	Monterey	S. Andreas
W100	23.89 ± 0.81 c	26.63 ± 0.05 a	17.76 ± 0.03 f
W80	23.38 ± 0.10 d	25.23 ± 0.01 b	16.90 ± 0.25 g
W60	23.72 ± 0.01 cd	22.18 ± 0.06 e	18.08 ± 0.12f
Folates (µg 100 g^{-1} of FW)			
W100	27.21 ± 0.74 de	28.87 ± 0.32 cd	29.94 bc ± 0.73 bc
W80	28.51 ± 1.17 cde	31.50 ± 0.67 ab	27.02 ± 2.38 e
W60	28.47 ± 0.65 cde	33.12 ± 0.19 a	30.78 ± 0.65 b
TPH (mgGA 100 g^{-1} of FW)			
W100	367.80 ± 27 ab	367.10 ± 3.58 ab	349.81 ± 22.69 b
W80	351.38 ± 14.67 b	358.72 ± 1.46 b	393.11 ± 31.77 a
W60	359.38 ± 7.64 b	391.15 ± 9.94 a	422.50 ± 33.81 a
ACY (mg PEL-3- GLU 100 g^{-1} of FW)			
W100	35.49 ± 4.68 b	36.24 ± 0.62 b	44.27 ± 3.93 a
W80	33.58 ± 2.52 b	33.82 ± 0.25 b	42.50 ± 5.50 a
W60	32.25 ± 1.32 b	34.06 ± 1.72 b	44.63± 5.86 a
TAC (mg TroloxEq 100 g^{-1} of FW)			
W100	509.13 ± 19.93 bc	552.06 ± 8.98 abc	506.66 ± 31.91 c
W80	539.86 ± 49.26 bc	559.79 ± 18.45 abc	557.57 ± 37.87 abc
W60	535.92 ± 36.63 bc	561.27 ± 42.33 ab	599.23 ± 13.12 a

Note: Values with the same lowercase letter for the same parameter were not statistically different in Fisher's LSD test ($p < 0.05$). Values are expressed as the means of one year (2019) ± standard deviation.

4. Conclusions

In conclusion, the present study analyzed the responses of three different remontant cultivars of strawberry in water stress conditions. Although the experiment takes into account only one-year cultivation cycle, the results obtained are interesting and confirmed the need of future trials regarding the optimization of the irrigation management. In our experiment, even though the production was not as we expected, independently by the treatment, these cultivars showed satisfying results for all the evaluated parameters when irrigation was reduced by 20%. A further reduction of up to 40% in water amount led to a significant decline in the plants' vegetative and productive parameters. If the amount of water administrated is reduced by 20%, about 226 m^3 of water per hectare per cultivation cycle can be saved. A water shortage increased the fruits' sugar content, firmness, IP, folate content, total phenolic content, and total antioxidant capacity and the content of some phenolic acids. In addition, the genotype had a consistent impact on the plant's performance. Independent of the treatment, "S. Andreas" exhibited the highest fruit weight and the most balanced taste, with 14.6 °Brix and 14.5 mEQ of NaOH 100 g^{-1} of FW. Nevertheless, our results suggest that "Monterey" is the most preferable remontant cultivar among the studied genotypes. In fact, for all the treatments, "Monterey" exhibited an appreciable firmness, around 420 g, indicating that this plant's fruits are most suitable for a longer shelf life on the market. "Monterey" also performed well in terms of commercial production (g/plant), showing a reduction in only about 4% at W80 compared to the production at W100. For this cultivar, these results are appreciable, especially when considering that "Monterey" achieved good values in terms of the contents of vitamin C

(26.63 mg 100 g^{-1} of FW at W100 and 25.23 mg 100 g^{-1} of FW at W80), folate (28.87 µg 100 g^{-1} of FW at W100 and 31.50 µg 100 g^{-1} of FW at W80), TAC (552.06 mg Trolox eq 100 g^{-1} of FW at W100 and 559.79 Trolox eq 100 g^{-1} of FW at W80), and phenolic acids (chlorogenic acid, caffeic acid, and ellagic acid were up to 40 mg 100 g^{-1} of FW for all treatments). This type of experiment is essential for identifying the cultivars most suitable for a specific environment. Nevertheless, to avoid climatic and environmental influences, a "closed" experimental site, with controlled cultivation conditions, is strongly suggested. Furthermore, nowadays, a specific protocol for water management is mandatory to ensure greater economic and environmental sustainability for high-quality strawberry production for consumers.

Supplementary Materials: The following supporting information can be downloaded at: https://www.mdpi.com/article/10.3390/horticulturae9091026/s1, Figure S1: Soil water potential calculated by the tensiometers; Figure S2: Temperature from July 2019 until October 2019 in the ASSAM experimental field; Figure S3: Experimental design of the trial.

Author Contributions: Conceptualization, B.M. and F.C.; methodology, M.M. and D.R.; software, M.M., D.R. and F.C.; validation, L.M., B.M. and F.C.; formal analysis, V.P., F.B., Y.A.D., D.R. and M.M.; investigation, M.M., D.R. and F.C.; resources, B.M. and F.C.; data curation, M.M., D.R. and F.C.; writing—original draft preparation,—D.R.; writing—review and editing, L.M. and B.M.; visualization, M.M. and D.R.; supervision, F.C. and B.M.; project administration, F.C.; funding acquisition, F.C. and B.M. All authors have read and agreed to the published version of the manuscript.

Funding: This research was funded by European Union's Horizon 2020 research and innovation program, grant number no. 679303 (GOODBERRY project).

Data Availability Statement: Publicly available datasets were analyzed in this study. This data can be found here: https://hdl.handle.net/11566/274632, accessed on 1 March 2023.

Conflicts of Interest: The authors declare no conflict of interest.

References

1. Ibrahim, M.H.; Nulit, R.; Sakimin, S.Z. Influence of Drought Stress on Growth, Biochemical Changes and Leaf Gas Exchange of Strawberry (*Fragaria* × *ananassa* Duch.) in Indonesia. *AIMS Agric. Food* **2022**, *7*, 37–60. [CrossRef]
2. Navarro-Hortal, M.D.; Romero-Márquez, J.M.; Esteban-Muñoz, A.; Sánchez-González, C.; Rivas-García, L.; Llopis, J.; Cianciosi, D.; Giampieri, F.; Sumalla-Cano, S.; Battino, M.; et al. Strawberry (*Fragaria* × *ananassa* cv. *Romina*) methanolic extract attenuates Alzheimer's beta amyloid production and oxidative stress by SKN-1/NRF and DAF-16/FOXO mediated mechanisms in *C. elegans*. *Food Chem.* **2022**, *372*, 131272. [CrossRef]
3. Battino, M.; Giampieri, F.; Cianciosi, D.; Ansary, J.; Chen, X.; Zhang, D.; Gil, E.; Forbes-Hernández, T. The roles of strawberry and honey phytochemicals on human health: A possible clue on the molecular mechanisms involved in the prevention of oxidative stress and inflammation. *Phytomedicine* **2021**, *86*, 153170. [CrossRef] [PubMed]
4. Iqbal, A.Z.; Javaid, N.; Hameeda, M. Synergic interactions between berry polyphenols and gut microbiota in cardiovascular diseases. *Med. J. Nutr. Metab.* **2022**, *15*, 555–573. [CrossRef]
5. Micek, A.; Owczarek, M.; Jurek, J.; Guerrera, I.; Torrisi, S.A.; Grosso, G.; Alshatwi, A.A.; Godos, J. Anthocyanin-rich fruits and mental health outcomes in an Italian cohort. *J. Berry Res.* **2022**, *14*, 551–564. [CrossRef]
6. Ullah, H.; De Filippis, A.; Santarcangelo, C.; Daglia, M. Epigenetic regulation by polyphenols in diabetes and related complications. *Med. J. Nutr. Metab.* **2020**, *13*, 289–310. [CrossRef]
7. Bhutia, P.O.; Kewlani, P.; Pandey, A.; Rawat, S.; Bhatt, I.D. Physico-chemical properties and nutritional composition of fruits of the wild Himalayan strawberry (*Fragaria nubicola* Lindle.) in different ripening stages. *J. Berry Res.* **2021**, *11*, 481–496. [CrossRef]
8. Battino, M.; Forbes-Hernández, T.Y.; Gasparrini, M.; Afrin, S.; Cianciosi, D.; Zhang, J.; Manna, P.P.; Reboredo-Rodríguez, P.; Varela Lopez, A.; Quiles, J.L.; et al. Relevance of functional foods in the Mediterranean diet: The role of olive oil, berries and honey in the prevention of cancer and cardiovascular diseases. *Crit. Rev. Food Sci. Nutr.* **2019**, *59*, 893–920. [CrossRef]
9. Galli, V.; da Silva Messias, R.; Perin, E.C.; Borowski, J.M.; Bamberg, A.L.; Rombaldi, C.V. Mild Salt Stress Improves Strawberry Fruit Quality. *LWT* **2016**, *73*, 693–699. [CrossRef]
10. Sperry, J.S.; Love, D.M. What Plant Hydraulics Can Tell Us about Responses to Climate-Change Droughts. *New Phytol.* **2015**, *207*, 14–27. [CrossRef]
11. FAO; UN Water. *Progress on the Level of Water Stress: Global Status and Acceleration Needs for SDG Indicator 6.4.2*; FAO: Rome, Italy; United Nations Water (UN Water): Rome, Italy, 2021. [CrossRef]

12. Tripathi, A.D.; Mishra, R.; Maurya, K.K.; Singh, R.B.; Wilson, D.W. Estimates for World Population and Global Food Availability for Global Health. In *The Role of Functional Food Security in Global Health*; Elsevier Inc.: Amsterdam, The Netherlands, 2018; Volume 106, pp. 3–24. [CrossRef]
13. Mancosu, N.; Snyder, R.L.; Kyriakakis, G.; Spano, D. Water Scarcity and Future Challenges for Food Production. *Water* **2015**, *7*, 975–992. [CrossRef]
14. Vanham, D.; Hoekstra, A.Y.; Wada, Y.; Bouraoui, F.; de Roo, A.; Mekonnen, M.M.; van de Bund, W.J.; Batelaan, O.; Pavelic, P.; Bastiaanssen, W.G.M.; et al. Physical Water Scarcity Metrics for Monitoring Progress towards SDG Target 6.4: An Evaluation of Indicator 6.4.2 "Level of Water Stress". *Sci. Total Environ.* **2018**, *613*, 218–232. [CrossRef] [PubMed]
15. UNWATER. Available online: https://sdg6data.org/en/country-or-area/Italy#anchor_6.4.2 (accessed on 10 December 2022).
16. ISAC-CNR. Available online: https://www.isac.cnr.it/ (accessed on 15 November 2022).
17. Van Halsema, G.E.; Vincent, L. Efficiency and Productivity Terms for Water Management: A Matter of Contextual Relativism versus General Absolutism. *Agric. Water Manag.* **2012**, *108*, 9–15. [CrossRef]
18. Li, P.; Karunanidhi, D.; Subramani, T.; Srinivasamoorthy, K. Sources and Consequences of Groundwater Contamination. *Arch. Environ. Contam. Toxicol.* **2021**, *80*, 1–10. [CrossRef] [PubMed]
19. Chaves, N.P.; Pocasangre, L.E.; Elango, F.; Rosales, F.E.; Sikora, R. Combining Endophytic Fungi and Bacteria for the Biocontrol of Radopholus Similis (Cobb) Thorne and for Effects on Plant Growth. *Sci. Hortic.* **2009**, *122*, 472–478. [CrossRef]
20. Flexas, J.; Bota, J.; Escalona, J.M.; Sampol, B.; Medrano, H. Effects of drought on photosynthesis in grapevines under field conditions: An evaluation of stomatal and mesophyll limitations. *Funct. Plant Biol.* **2002**, *29*, 461–471. [CrossRef]
21. Ariza, M.T.; Miranda, L.; Gómez-Mora, J.A.; Medina, J.J.; Lozano, D.; Gavilán, P.; Soria, C.; Martínez-Ferri, E. Yield and Fruit Quality of Strawberry Cultivars under Different Irrigation Regimes. *Agronomy* **2021**, *11*, 261. [CrossRef]
22. Sabbadini, S.; Capocasa, F.; Battino, M.; Mazzoni, L.; Mezzetti, B. Improved Nutritional Quality in Fruit Tree Species through Traditional and Biotechnological Approaches. *Trends Food Sci. Technol.* **2021**, *117*, 125–138. [CrossRef]
23. Strawberry Breeding & Research. University of California, Davis. Available online: https://research.ucdavis.edu/industry/ia/industry/strawberry/cultivars/ (accessed on 5 October 2022).
24. Marche Region Directive 786 on 10 July 2017. Available online: https://www.norme.marche.it/Delibere/2017/DGR0786_17.pdf (accessed on 16 January 2019).
25. Davey, B.G.; Conyers, M.K. Determining the pH of acid soils. *Soil Sci.* **1988**, *146*, 141–150. [CrossRef]
26. Gee, G.W.; Bauder, J.W. Particle-size analysis. In *Methods of Soil Analysis: Part 1 Physical and Mineralogical Methods*; American Society of Agronomy Inc.: Washington, DC, USA, 1986; Volume 5, pp. 383–411.
27. Olsen, S.R. *Estimation of Available Phosphorus in Soils by Extraction with Sodium Bicarbonate*; (No. 939); US Department of Agriculture: Washington, DC, USA, 1954.
28. McKeague, J.A.; Brydon, J.E.; Miles, N.M. Differentiation of forms of extractable iron and aluminum in soils. *Soil Sci. Soc. Am. J.* **1971**, *35*, 33–38. [CrossRef]
29. Barbafieri, M.; Lubrano, L.; Petruzzelli, G. Characterization of heavy metal pollution in soil. *Ann. Chim.* **1996**, *86*, 635–652.
30. Jeffrey, A.J.; McCallum, L.E. Investigation of a hot 0.01 M CaCl2 soil boron extraction procedure followed by ICP-AES analysis. *Commun. Soil Sci. Plant Anal.* **1988**, *19*, 663–673. [CrossRef]
31. Bremner, J.M.; Keeney, D.R. Determination and isotope-ratio analysis of different forms of nitrogen in soils: 3. Exchangeable ammonium, nitrate, and nitrite by extraction-distillation methods. *Soil Sci. Soc. Am. J.* **1966**, *30*, 577–582. [CrossRef]
32. Knudsen, D.; Peterson, G.A.; Pratt, P.F. Lithium, sodium, and potassium. In *Methods of Soil Analysis: Part 2 Chemical and Microbiological Properties*; American Society of Agronomy Inc.: Washington, DC, USA, 1983; Volume 9, pp. 225–246.
33. Capocasa, F.; Balducci, F.; Di Vittori, L.; Mazzoni, L.; Stewart, D.; Williams, S.; Hargreaves, R.; Bernardini, D.; Danesi, L.; Zhong, C.F.; et al. Romina and Cristina: Two New Strawberry Cultivars with High Sensorial and Nutritional Values. *Int. J. Fruit Sci.* **2016**, *16*, 207–219. [CrossRef]
34. Marcellini, M.; Mazzoni, L.; Raffaelli, D.; Pergolotti, V.; Balducci, F.; Capocasa, F.; Mezzetti, B. Evaluation of Single-Cropping under Reduced Water Supply in Strawberry Cultivation. *Agronomy* **2022**, *12*, 1396. [CrossRef]
35. Sadler, G.D.; Murphy, P.A. pH and titratable acidity. *Food Anal.* **2010**, *4*, 219–238.
36. Mezzetti, B.; Balducci, F.; Capocasa, F.; Zhong, C.F.; Cappelletti, R.; Di Vittori, L.; Mazzoni, L.; Giampieri, F.; Battino, M. Breeding Strawberry for Higher Phytochemicals Content and Claim It: Is It Possible? *Int. J. Fruit Sci.* **2016**, *16*, 194–206. [CrossRef]
37. Iniesta, M.D.; Perez-Conesa, D.; Garcia-Alonso, J.; Ros, G.; Periago, M.J. Folate content in tomato (*Lycopersicon esculentum*): Influence of cultivar, ripeness, year of harvest, and pasteurization and storage temperatures. *J. Agric. Food Chem.* **2009**, *57*, 4739–4745. [CrossRef] [PubMed]
38. Jastrebova, J.; Witthoft, C.; Grahn, A.; Svensson, U.; Jagerstad, M. HPLC determination of folates in raw and processed beetroots. *Food Chem.* **2003**, *80*, 579–588. [CrossRef]
39. Strålsjö, L.; Arkbåge, K.; Witthöft, C.; Jägerstad, M. Evaluation of a radioprotein-binding assay (RPBA) for folate analysis in berries and milk. *Food Chem.* **2002**, *79*, 525–534. [CrossRef]
40. Fredericks, C.H.; Fanning, K.J.; Gidley, M.J.; Netzel, G.; Zabaras, D.; Herrington, M.; Netzel, M. High-Anthocyanin Strawberries through Cultivar Selection. *J. Sci. Food Agric.* **2013**, *93*, 846–852. [CrossRef]
41. Giusti, M.; Wrolstad, R.E. Characterization and Measurement of anthocyanins by UV-visible Spectroscopy. In *Current Protocols in Food Analytical Chemistry*, 1st ed.; Wrolstad, R.E., Ed.; John Wiley & Sons: New York, NY, USA, 2001; pp. F1.2.1–F1.2.13.

42. Miller, N.J.; Rice-Evans, C.; Davis, M.J. A novel method for measuring antioxidant capacity and its application to monitoring the antioxidant status in premature neonates. *Clin. Sci.* **1993**, *83*, 407–412. [CrossRef]
43. Re, R.; Pellegrini, N.; Proteggente, A.; Pannala, A.; Yang, M. Antioxidant activity applying an improved ABTS radical cation decolorization assay. *Free Radic. Biol. Med.* **1999**, *26*, 1231–1237. [CrossRef]
44. Slinkard, K.; Singleton, V.L. Total phenol analysis: Automation and comparison with manual methods. *Am. J. Enol. Vitic.* **1977**, *28*, 49–55. [CrossRef]
45. Gehrmann, H. Growth, Yield and Fruit Quality of Strawberries As Affected By Water Supply. *Acta Hortic.* **1985**, *171*, 463–469. [CrossRef]
46. Awang, Y.B.; Atherton, J.G.; Taylor, A.J. Salinity Effects on Strawberry Plants Grown in Rockwool. I. Growth and Leaf Water Relations. *J. Hortic. Sci.* **1993**, *68*, 783–790. [CrossRef]
47. Yuan, B.Z.; Sun, J.; Nishiyama, S. Effect of Drip Irrigation on Strawberry Growth and Yield inside a Plastic Greenhouse. *Biosyst. Eng.* **2004**, *87*, 237–245. [CrossRef]
48. Grant, O.M.; Davies, M.J.; Johnson, A.W.; Simpson, D.W. Physiological and Growth Responses to Water Deficits in Cultivated Strawberry (*Fragaria* × *ananassa*) and in One of Its Progenitors, *Fragaria chiloensis*. *Environ. Exp. Bot.* **2012**, *83*, 23–32. [CrossRef]
49. Kumar, P.S.; Choudhary, V.K.; Bhagawati, R. Influence of Mulching and Irrigation Level on Water-Use Efficiency, Plant Growth and Quality of Strawberry (*Fragaria* × *ananassa*). *Indian J. Agric. Sci.* **2012**, *82*, 127–133.
50. Saied, A.S.; Keutgen, A.J.; Noga, G. The Influence of NaCl Salinity on Growth, Yield and Fruit Quality of Strawberry Cvs. "Elsanta" and "Korona". *Sci. Hortic.* **2005**, *103*, 289–303. [CrossRef]
51. Terry, L.A.; Chope, G.A.; Giné Bordonaba, J. Effect of Water Deficit Irrigation and Inoculation with *Botrytis Cinerea* on Strawberry (*Fragaria* × *ananassa*) Fruit Quality. *J. Agric. Food Chem.* **2007**, *55*, 10812–10819. [CrossRef] [PubMed]
52. Giné Bordonaba, J.; Terry, L.A. Manipulating the Taste-Related Composition of Strawberry Fruits (*Fragaria* × *ananassa*) from Different Cultivars Using Deficit Irrigation. *Food Chem.* **2010**, *122*, 1020–1026. [CrossRef]
53. Martínez-Ferri, E.; Soria, C.; Ariza, M.T.; Medina, J.J.; Miranda, L.; Domíguez, P.; Muriel, J.L. Water Relations, Growth and Physiological Response of Seven Strawberry Cultivars (*Fragaria* × *ananassa* Duch.) to Different Water Availability. *Agric. Water Manag.* **2016**, *164*, 73–82. [CrossRef]
54. Cardeñosa, V.; Medrano, E.; Lorenzo, P.; Sánchez-Guerrero, M.C.; Cuevas, F.; Pradas, I.; Moreno-Rojas, J.M. Effects of Salinity and Nitrogen Supply on the Quality and Health-Related Compounds of Strawberry Fruits (*Fragaria* × *ananassa* Cv. Primoris). *J. Sci. Food Agric.* **2015**, *95*, 2924–2930. [CrossRef]

horticulturae

Article

Fruit Quality of Several Strawberry Cultivars during the Harvest Season under High Tunnel and Open Field Environments

Hiral Patel [1], Toktam Taghavi [1,*] and Jayesh B. Samtani [2,*]

[1] Agricultural Research Station, College of Agriculture, Virginia State University, Petersburg, VA 23806, USA; patelhiral712@gmail.com

[2] Hampton Roads Agricultural Research and Extension Center, School of Plant and Environmental Sciences, Virginia Tech, Virginia Beach, VA 23455, USA

* Correspondence: ttaghavi@vsu.edu (T.T.); jsamtani@vt.edu (J.B.S.)

Abstract: Parameters such as titratable acids (TA), total soluble solids (TSS), and their ratio (TSS/TA) are critical in determining strawberry fruit quality and the value of new cultivars. Ten strawberry cultivars were evaluated in two environments (open field and high tunnel) in the city of Virginia Beach. The objective was to evaluate the fruit quality characteristics (total soluble solids, titratable acidity TA, and total anthocyanin content) of newer strawberry cultivars grown in the annual hill plasticulture systems in coastal Virginia climatic conditions. Another objective was to measure the correlation between TA and a new digital meter (pocket acidity meter; PAM). Fruits were harvested weekly and TSS was measured using a refractometer. Acidity was measured using the pocket acidity meter and titratable acidity by a single sample titrimeter. Genetics significantly affected total anthocyanin content, TSS, TA, and acidity. The effect of the environments (high tunnel and open field) was not significant on TSS but significant on total anthocyanin content, TA, and acidity. "Flavorfest" had the highest and "Sweet Ann" the lowest anthocyanin content, TSS, and TA among the cultivars. The acidity (PAM data) showed a different level of correlation than TA, with a higher correlation for the open field than the high tunnel. On average, when outliers were removed, there was a regression of TA = 2.22(PAM) + 0.49 between the two data sets, with $R^2 = 0.47$.

Keywords: flavonoids; pigments; protected environment; titratable acidity; acidity; total soluble solids

Citation: Patel, H.; Taghavi, T.; Samtani, J.B. Fruit Quality of Several Strawberry Cultivars during the Harvest Season under High Tunnel and Open Field Environments. *Horticulturae* **2023**, *9*, 1084. https://doi.org/10.3390/horticulturae9101084

Academic Editor: Giovanni Battista Tornielli

Received: 29 July 2023
Revised: 18 September 2023
Accepted: 26 September 2023
Published: 28 September 2023

1. Introduction

Strawberries (*Fragaria × ananassa*) are planted for their usually red, sweet, and aromatic fruit. However, due to their short shelf life, worldwide strawberry production is relatively low compared with other fruit crops. Global strawberry production reached 9.22 million metric tons in 2017 [1]. The annual strawberry production in the United States was 1.43 million tons in 2018, accounting for a third of the total world's production, leading the global strawberry market [2,3].

Favorable climate conditions make the states of California and Florida the largest producers of strawberries in the United States [1]. Outside of California and Florida, the South Atlantic is the most productive region (including Virginia) for fresh market strawberries in the United States. There is increasing interest in commercial strawberry production for local markets in Virginia and surrounding states. This region has about 1000 hectares of strawberries [4,5].

Strawberry cultivars are more sensitive to environmental conditions than other fruit crops and their anatomy, morphology, growth habits, and reproductive growth have been studied [6]. However, the relative contribution of genotypic and environmental conditions to fruit quality is not well studied. Additionally, it is unknown to what extent these fruit quality traits remain stable throughout the changing environmental conditions of the cropping season [7].

A high tunnel is a semi-permanent structure with plastic covers that uses passive ventilation for temperature and air humidity adjustment and modifies the environmental conditions to improve fruit quality, extend the season, and protect plants from extreme weather conditions. Strawberries have been cultured in high tunnels or open fields under mild winter climates for out-of-season production. In these cases, specific treatments may secure better performance depending on the cultivar and the environmental conditions [8]. It is known that changes in environmental conditions affect the fruiting capacity and fruit quality, among other characteristics, and can be a source of stress for the crop depending on the ability of the cultivars to cope with it [9]. In this sense, antioxidant compounds of a polyphenolic nature (i.e., anthocyanins) play an essential role in the general mechanisms of the response to different stressors; therefore, changes in environmental conditions could be expected to influence the composition and synthesis of these compounds in fruits. Therefore, different environmental conditions (open field vs. high tunnel) can affect the quality of the fruits in different strawberry genotypes and, consequently, their acceptance by the consumers [7].

Strawberries contain several bioactive phytochemicals, including anthocyanins, flavonols, and phenolic acids. Anthocyanins are a group of flavonoid pigments that are responsible for a wide range of red colors in fruits. Anthocyanins in berries are partly responsible for the high antioxidant activity and have demonstrated a role in protecting plant and human health [10].

Consumers prefer strawberries with a wide range of sensory features. The quality components can be sensory and nutritional [10]. Parameters such as titratable acids (TA), total soluble solids (TSS), and their ratio (TSS/TA) are critical in determining strawberry fruit quality and the value of new cultivars. The TSS to TA ratio is critical in evaluating fruit quality because it determines flavor harmony. Hence, along with fruit color, they are significant factors in determining strawberry fruit quality [10].

High TSS and TA contents represent general selection criteria for the flavor of strawberry fruits in breeding programs. A good and well-balanced flavor for strawberries is based on a high sugar and a comparatively high acid content (i.e., the balance between sweetness and acidity). Their ratio (TSS/TA) is commonly used to evaluate the taste and ripening stage of the fruit. A ratio of (TSS/TA) of 8.5–14 is considered an appropriate balance of sweet–tart flavor notes in strawberries for human palatability [11–13]. In another report, the minimum TSS and maximum TA levels for an acceptable flavor of strawberry were recommended to be 7% and 0.8%, respectively [14,15]. In a third report, the results showed that eating quality was more strongly related to TSS than to TA and a higher TA than recommended (even close to 1%) was still acceptable if combined with high TSS [15].

There are several methods to measure acidity. The pH scale is used to measure the hydrogen ion concentration of a solution. However, pH measurements are not always accurate, especially when dealing with complex or heterogeneous solutions such as fruit extracts.

Titratable acidity measures the hydrogen ions by neutralizing them with sodium hydroxide (base) in a known sample quantity. The amount of the base needed for neutralization reflects the acid content. Compared with pH, titratable acidity is a better predictor of sourness and more closely related to the taste. The process of measuring TA is tedious and labor intensive, even when an automated titrimeter is used, and it needs large amounts of costly reagents, technical expertise, instruments, and a laboratory. A digital meter was recently introduced that quickly measures acidity with minimal preparation and no reagents (pocket acidity meter, PAM F5, Atago, Japan). The PAM measures the acidity level through the electro-conductivity method using electrical current. An electrical current passes through a solution via ions. The conductivity of a solution depends on the concentration of all the ions present and their mobility; however, hydrogens ion are the major contributor to the electrical conductivity of a solution due to their lightweight and faster mobility (on average 10 times more than other ions).

In a previous study, several small fruits' TA and acidity (PAM) levels were measured with an automated titrimeter and a digital meter. A regression was conducted between the two values [16] and the data suggested a strong correlation for blackberries and blueberries, with R^2 values of 0.82 and 0.85, respectively. The correlations were not as strong for raspberries and strawberries and there were some inconsistencies in the data. The reason for the inconsistency was unclear; the R^2 values were 0.53 and 0.25 for red raspberry and strawberry, respectively. For the first objective, we expanded the experiment to include more strawberry cultivars and two different environments (high tunnel and open field). The objective was to determine whether the acidity (PAM data) provides a more reliable and precise method to determine the acidity of strawberry extract than titratable acidity.

The second objective of this research was to evaluate fruit quality characteristics (TSS, TA, and total anthocyanin content) of newer strawberry cultivars during harvest season in the annual hill plasticulture system in an open field and under high-tunnel conditions in coastal Virginia climatic conditions (USDA Plant Hardiness Zones 7 and 8). The goal was to determine the effect of the environments and genotypes on strawberry fruit quality during the cropping season.

2. Materials and Methods

Strawberries were planted from September 2019 through June 2020 at Hampton Roads Agricultural Research and Extension Center (AREC) in the city of Virginia Beach, Virginia ($36°9'$ N, $76°2'$ W). Strawberries were planted in a Randomized Complete Block Design (RCBD) in four replicates. Soil samples from the top 20 cm were collected before the experiment in both environments and sent to the Virginia Polytechnic Institute and State University Soil Testing Lab. The soil was a tetotum loam with a pH of 5.9 and limestone was broadcast on 5 September 2019, at 1120 kg ha^{-1}. The soil pH was adjusted to the desired level of 6.2. The drip irrigation and irrigation systems were set up using a 0.38 mm single drip line with a 30.5 cm emitter spacing (Berry Hill Irrigation, Inc., Buffalo Junction, VA, USA). Fertilizers were applied with pre-plant fertilizer at 69 kg ha^{-1} nitrogen, using Nutrisphere-N (N-P-K ratio of 34-0-0, Southern States Cooperatives Inc., Chesapeake, VA, USA).

A total of ten cultivars ("Camino Real", "Chandler", "Merced", 'Rocco", "Ruby June", "Albion", "Flavorfest", "Keepsake", "San Andreas", and "Sweet Ann") were evaluated in two environments, high tunnel and open field, in a randomized complete block design with four replicates. Short-day cultivars were "Chandler", "Merced", "Camino Real" (all from the University of California, Davis), "Flavorfest" and "Keepsake" (USDA, Beltsville, MD, USA), "Rocco" (North Carolina State University), and "Ruby June" (Lassen Canyon Nursery). Day-neutral cultivars were "Albion", "San Andreas" (UC Davis), and "Sweet Ann" (Lassen Canyon Nursery). Strawberry plugs of all cultivars were ordered from Aaron's Creek Farms Plant Nursery, Buffalo Junction, VA, USA. Fruits were harvested weekly from 2 April until 1 June (Figure 1).

Strawberries were frozen and transferred to the Postharvest Research Lab at Virginia State University (VSU), kept frozen on ice packs during transportation, and placed at $-32\,°C$ once they arrived at VSU until further use.

Half of the frozen strawberries were thawed at room temperature, and the juice was used to measure the fruit quality parameters. TheTSS content was measured by a refractometer (Atago, Tokyo, Japan). A few drops of the juice were placed on the refractometer, and the data were presented as °Brix. Acidity was measured by the PAM F5 (pocket acidity meter) from Atago (Tokyo, Japan) and titratable acidity by a single sample titrimeter (EasypH Mettler Toledo, Greifensee, Switzerland). An aliquot of juice or puree was used to prepare a 1:50 solution by adding 1 mL of juice and 49 mL (or g) of distilled deionized water and the solution was mixed. A 0.2 to 0.5 mL aliquot was placed on the refractometer (precalibrated with a 0.04% citric acid solution) using mode 4 (strawberry) setting. The remaining solution was titrated to an endpoint of 8.2 using 0.1 N sodium hydroxide using titrimeters. The TA was calculated based on % citric acid equivalents. Titrimeter readings were plotted against PAM readings, and the linear least squares that fitted the equation with R^2 were calculated.

High tunnel   Open field

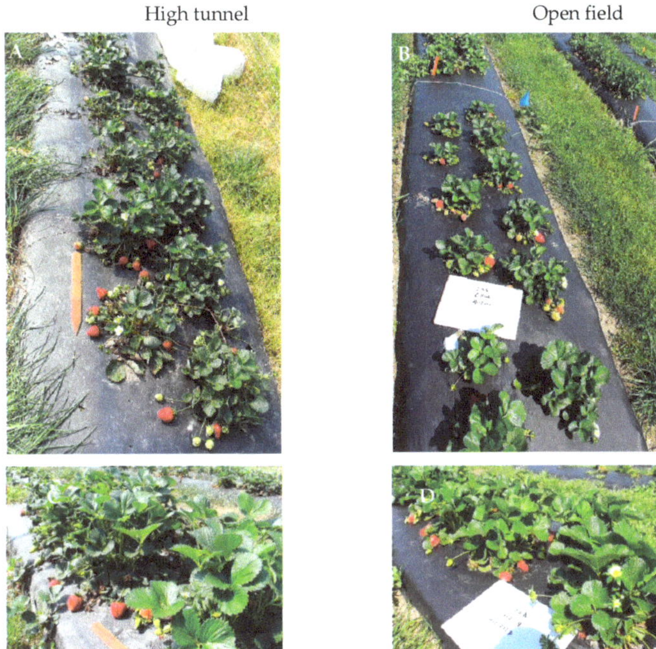

Figure 1. Top row, "Camino Real" plants during the harvest season in high-tunnel (**A**) and open-field environments (**B**). Bottom row, "Ruby June" plants during the harvest season in high-tunnel (**C**) and open-field environments (**D**).

Half of the frozen strawberries were sliced and one gram was freeze-dried at −80 °C (VirTis Freezemobile freeze dryers SP Scientific, Warminster, PA, USA) for anthocyanin extraction. Freeze-dried samples were ground to a fine powder, and anthocyanins were extracted using acidified methanol. Extracts were obtained by adding 20 mL of methanol (acidified with 0.01% HCl) to the strawberry powder. The homogenates were incubated at 4 °C in the dark on a shaker for 24 h. At the end of the incubation period, the homogenates were centrifuged at 4 °C and 7000 rpm for 15 min. The supernatant was then removed and the absorbance of anthocyanin measured using the spectrophotometer at 530 and 657 nm [17]. The anthocyanin concentration was calculated by the following formula and given as A/g fresh fruit tissue (A/gFW), where A = absorbance at 530 and 657 nm, V = volume of extract (mL), and M = fresh mass of the sample (g).

$$Total\ Anthocyanins = \frac{A_{530} - 0.3\ A_{657} \times V}{M} \tag{1}$$

Experimental Design and Statistical Analysis

Fruit quality data were analyzed using Proc GLM of Statistical Analysis Software Version 9.4 [18]. For the regression analysis, the Proc REG procedure was used. The data from four replicates were averaged for each experimental unit. The data were averaged for the harvest season to calculate the means for environments or cultivars.

3. Results and Discussion

3.1. Total Anthocyanins

The total anthocyanin content averaged over the harvest season was lowest in "Sweet Ann" and "Ruby June" (142 and 152 A/gFW, respectively) and highest in "Flavorfest", followed by "Chandler" (289 and 254 A/gFW, respectively) (Table 1). The total anthocyanin was also affected by the environment and was 23% higher in strawberries grown in the high tunnel than in the open field (Table 1).

Table 1. Total anthocyanin content over the growing season of different strawberry cultivars grown in the open field and high tunnel.

Cultivar	Total Anthocyanin Content (A/gFW)	
Flavorfest	289.4	a *
Keepsake	184.0	d
Rocco	222.4	bc
Albion	196.5	cd
Ruby June	151.8	e
Merced	187.2	cd
Chandler	254.1	b
Camino Real	219.9	c
Sweet Ann	141.7	e
San Andreas	212.2	cd
LSD	32.0	
Environment		
High tunnel	246.5	a
Open field	197.7	b
LSD	16.6	

Notes: * Means with the same letters within a column are not significantly different at $p = 0.05$.

When the interaction of cultivars and the environments were analyzed, "Flavorfest" grown in the high tunnel had the highest total anthocyanin content (300 A/gFW), which was 127% higher than "Sweet Ann" (132 A/gFW) in the open field (Table 2). A higher anthocyanin content has been reported for strawberries and raspberries grown in the high tunnel compared with those grown in the open field [19,20].

Table 2. The interaction of environment and cultivar on the total anthocyanin content (A/gFW) of strawberries.

Environment	Cultivar	Total Anthocyanin Content (A/gFW)	Standard Error
High Tunnel	Flavorfest	300.0 a *	10.16
	Keepsake	212.7 bcd	19.80
	Rocco	188.9 bcd	23.66
	Ruby June	165.3 cd	25.56
	Merced	172.9 cd	25.56
	Chandler	193.7 bcd	31.30
	Sweet Ann	211.7 bcd	31.30
	San Andreas	228.7 abc	62.60
	Avg	209.3	
Open Field	Flavorfest	208.8 bcd	28.00
	Keepsake	161.9 cd	17.36
	Rocco	229.7 abc	11.07
	Albion	196.5 bcd	11.63
	Ruby June	148.6 cd	12.52
	Merced	190.1 bcd	11.63
	Chandler	263.4 ab	12.28
	Camino Real	219.9 bc	11.63
	Sweet Ann	132.0 d	11.63
	San Andreas	211.6 bcd	11.63
	Avg	196.2	

Notes: * Means with the same letters within a column are not significantly different at $p = 0.05$. Fruits for cultivars "Albion" and "Camino Real" in the high tunnel were not available for analysis.

3.2. Total Soluble Solids

Total soluble solids (TSS) content was highest in "Flavorfest" (9.65 °Brix) and "Keepsake" and lowest in "San Andreas" (7.16 °Brix) when averaged over the two environments (Table 3). The total soluble solids content for "Rocco", "Albion", and "Ruby June" were not significantly different than "Flavorfest" and "Rocco". TSS content was not significantly different between the strawberries grown in the high tunnel and open field; both were above 8 (°Brix) when averaged over the cultivars.

Table 3. The total soluble solids (TSS) content, acidity, and titratable acidity (TA) of strawberry cultivars averaged over the harvest season and in the high-tunnel or open-field environments.

Cultivar	TSS (°Brix)	Acidity (%, PAM)	TA (%)	TSS/TA
Flavorfest	9.65 a *	0.67 a	1.94 a	4.97
Keepsake	9.20 a	0.63 ab	1.79 ab	5.14
Rocco	8.83 ab	0.63 ab	1.93 a	4.58
Albion	8.82 ab	0.58 cd	1.77 abc	4.98
Ruby June	8.79 ab	0.60 bc	1.77 abc	4.97
Merced	8.22 bc	0.51 e	1.52 d	5.41
Chandler	7.93 bcd	0.57 cd	1.79 ab	4.43
Camino Real	7.59 cd	0.49 e	1.60 cd	4.74
Sweet Ann	7.56 cd	0.51 e	1.54 d	4.91
San Andreas	7.16 d	0.55 d	1.64 bcd	4.37
LSD	0.87	0.04	0.16	
Environment				
High Tunnel	8.08 a	0.62 a	1.76 a	4.59
Open Field	8.30 a	0.50 b	1.64 b	5.06
LSD	0.32	0.14	0.06	

* Means with the same letters within a column are not significantly different at $p = 0.05$.

The TSS values in this study are slightly lower than previous reports but within the acceptable range of 7–12% reported by the Oregon Strawberry Commission [11] and Keutgen and Pawelzik [12]. For example, "Korona" and "Elsanta" had TSS contents of 9.5 and 8.4%, respectively [14].

3.3. Acidity and Titratable Acidity

The acidity (PAM) values were lower than the TA values measured using the titrimeter. The PAM values were highest in "Flavorfest", "Keepsake", and "Rocco" and lowest in "Sweet Ann", "Merced''', and "Camino Real". The TA showed the same pattern: highest in "Flavorfest" and "Rocco" and lowest in "Sweet Ann" and "Merced" (Table 3). The differences among cultivars were subtle for TA values. Both values were higher in strawberries grown in the high tunnel than in the open field.

Genotype had substantial effects on the TSS and TA. Zhang et al. [21] stated that both TSS and TA are strongly influenced by genotype and that the TSS content was particularly high in "Sabrina", "Rubygem", "Sabrosa", and "Camarosa". Herrington et al. [22] and Saraçoğlu [23] found that altitudes had a weak effect on TSS. Gündüz and Özbay [13] also reported that genotype substantially affected the TSS and TA but that location/altitude had a weak effect on TSS. Similar results were reported by Andreotti et al. [24] for different altitudes in South Tyrol (Italy). Herrington et al. [22] found that the genotype effect was more significant than the growing location for TA. Gündüz et al. [13] reported that the TA content in strawberries varied more intensely due to fruit maturity, genotype, and nutrition than ecological factors. Our data also confirm that the TSS varied more intensely due to the genotype than due to the open-field or high-tunnel environments. Changes in TA and acidity (PAM data) were more profound in different cultivars than in the outdoor or high-tunnel environments.

3.4. TSS/TA Ratio

The TSS/TA ratio is generally recommended as a quick measure of consumer acceptance [12]. The ratio varied from 4.4 to 5.4 in different cultivars and environments. The ratios in this experiment were lower than those previously reported, mainly due to the higher TA values [12,15]; however, these ratios did not correspond with any off-flavor taste.

The recommended TA is a maximum of 0.8%, whereas the minimum recommended TSS is 7% for an acceptable flavor in strawberries [14]. However, it was reported that eating quality was more strongly related to TSS than to TA [25]. A higher TA than recommended (even close to 1%) was still acceptable if combined with a high TSS content. Even so, an average sugar/acid ratio of 5.3 for the "Oso Grande", "Toyonoka", and "Mazi" cultivars was adequate to achieve the best quality [26]. Our results showed higher TA levels (close to 1) and a lower TSS/TA ratio than the recommendation [14] but with an acceptable strawberry flavor. Our data also confirmed that the TSS content has a more profound effect on eating quality than the TA.

3.5. TSS, Acidity, and TA over the Harvest Season

The TSS in strawberries did not change between the high-tunnel and open-field environments during the harvest season, except on the second harvest date, which was higher under the high tunnel than in the open field (Table 4). "Flavorfest" had a higher TSS content than the other cultivars, followed by "Keepsake". "San Andreas", "Sweet Ann", "Camino Real", and "Chandler" are among those that have the lowest TSS contents over the harvest season.

Table 4. Total soluble solids (TSS) of strawberry cultivars during the first five harvest dates and in high-tunnel or open-field environments.

Cultivar	°Brix-Avg	°Brix-Harvest1	°Brix-Harvest2	°Brix-Harvest3	°Brix-Harvest4	°Brix-Harvest5
Flavorfest	9.98 a *	9.76 a	9.10 a	10.13 a	- [b]	-
Keepsake	9.22 b	8.93 b	8.90 a	9.28 b	11.23 a	-
Rocco	8.80 b	8.55 bc	8.35 b	8.23 cd	8.63 b	8.79 ab
Albion	8.70 bc	7.30 e	7.48 c	8.47 bc	7.93 cd	8.53 b
Ruby June	8.79 b	8.15 cd	7.49 c	9.31 b	8.89 b	9.28 a
Merced	8.21 cd	8.12 cd	7.60 c	8.09 d	8.41 bc	8.68 b
Chandler	7.94 de	7.84 d	7.65 c	7.32 e	7.23 e	8.51 b
Camino Real	7.70 de	7.27 e	7.20 c	7.31 e	7.89 cd	7.76 c
Sweet Ann	7.59 ef	7.08 e	7.27 c	7.88 de	7.68 de	7.47 c
San Andreas	7.14 f	7.06 e	7.11 c	6.61 f	7.78 de	7.32 c
LSD	0.52	0.44	0.50	0.61	0.59	0.52
Environment						
High Tunnel	8.42 a	7.99 a	7.88 a	8.16 a	8.13 a	8.34 a
Open Field	8.30 a	7.93 a	7.56 b	8.03 a	8.11 a	8.24 a
LSD	0.23 ns	0.20 ns	0.20 **	0.22 ns	0.26 ns	0.25 ns

Notes: * Means with the same letters within a column are not significantly different at $p = 0.05$. [b] No data were collected during those weeks as the season for certain cultivars ended. ns means not significant and ** means significant at $p = 0.01$.

Titratable acidity and acidity (PAM data) were consistently higher in the high tunnel than in the open field strawberries during the harvest season (Tables 5 and 6). The TA value was highest in "Flavorfest" on the first harvest date; however, as we moved through the season, "Keepsake" and "Rocco" exhibited higher TA values. "Merced" had the lowest TA value throughout the harvest season (Table 6). Acidity (PAM) followed the same pattern, with "Flavorfest" having higher acidity and being replaced by "Keepsake" as we moved through the harvest season. The lowest acidities were measured in "Camino Real", "Merced", and "Sweet Ann" throughout the harvest season (Table 5).

Table 5. Acidity (%) of strawberry cultivars during the first five harvests and in high-tunnel or open-field environments (PAM data).

Cultivar	Acidity-Avg	Acidity-Harvest1	Acidity-Harvest2	Acidity-Harvest3	Acidity-Harvest4	Acidity-Harvest5
Flavorfest	0.71 a *	0.72 a	0.55 c	0.65 a	- [b]	-
Keepsake	0.64 b	0.69 a	0.62 ab	0.56 bc	0.71 a	-
Rocco	0.63 b	0.63 b	0.67 a	0.58 b	0.61 b	0.58 b
Albion	0.59 c	0.56 cde	0.55 c	0.60 ab	0.56 bc	0.59 ab
Ruby June	0.59 c	0.60 bcd	0.58 bc	0.61 ab	0.55 c	0.63 a
Merced	0.51 fg	0.53 ef	0.47 d	0.50 cd	0.45 de	0.53 c
Chandler	0.56 cd	0.60 bc	0.59 bc	0.58 b	0.53 cd	0.50 c
Camino Real	0.48 g	0.49 f	0.44 d	0.48 d	0.47 e	0.50 c
Sweet Ann	0.53 ef	0.54 def	0.54 c	0.51 cd	0.44 e	0.51 c
San Andreas	0.55 de	0.57 cde	0.56 bc	0.55 bcd	0.56 bc	0.51 c
LSD	0.03	0.05	0.06	0.06	0.05	0.04
Environment						
High Tunnel	0.65 a	0.65 a	0.64 a	0.63 a	0.62 a	0.63 a
Open Field	0.50 b	0.52 b	0.48 b	0.48 b	0.45 b	0.47 b
LSD	0.01	0.02	0.02	0.02	0.02	0.02

Notes: * Means with the same letters within a column are not significantly different at $p = 0.05$. [b] No data were collected during those weeks as the season for certain cultivars ended.

Table 6. The titratable acidity (TA, %) of strawberry cultivars during the first five harvests and in high-tunnel or open-field environments.

Cultivar	TA-Avg	TA-Harvest1	TA-Harvest2	TA-Harvest3	TA-Harvest4	TA-Harvest5
Flavorfest	2.04 a *	2.00 a	1.85 ab	1.85 ab	- [b]	-
Keepsake	1.80 c	1.91 ab	1.68 bc	1.80 ab	1.68 bc	-
Rocco	1.93 b	1.86 abc	1.97 a	1.74 abc	2.10 a	1.81 a
Albion	1.80 c	1.85 abc	1.77 ab	1.85 ab	1.52 bcd	1.82 a
Ruby June	1.77 c	1.67 bcd	1.85 ab	2.04 a	1.59 bcd	1.71 ab
Merced	1.52 e	1.62 cd	1.47 c	1.36 c	1.59 bcd	1.53 bc
Chandler	1.78 c	1.95 a	1.77 ab	1.70 abc	1.72 b	1.60 abc
Camino Real	1.60 de	1.69 bcd	1.68 bc	1.73 abc	1.37 cd	1.36 c
Sweet Ann	1.57 de	1.57 d	1.70 abc	1.53 bc	1.31 d	1.63 ab
San Andreas	1.65 d	1.75 abcd	1.82 ab	1.52 bc	1.76 b	1.60 abc
LSD	0.10	0.22	0.24	0.34	0.31	0.24
Environment						
High Tunnel	1.83 a	1.84	1.86 a	1.75 a	1.70 a	1.74 a
Open Field	1.64 b	1.72	1.64 b	1.64 a	1.56 b	1.54 b
LSD	0.04	0.10	0.10	0.12 ns	0.13	0.12

Notes: * Means with the same letters within a column are not significantly different at $p = 0.05$. [b] No data were collected during those weeks as the season for certain cultivars ended.

3.6. TA and Acidity (PAM Data) Regression

Titratable acidity estimates the sourness and sweetness of a fruit, but its measurement is labor intensive, expensive, and tedious, even when an automated titrimeter is used. Therefore, a pocket-sized digital meter (PAM) that quickly measures TA with minimal preparation was trialed for different cultivars of strawberries in two environments: high tunnel and open field. The regression data (TA and acidity) averaged over different cultivars and the two environments (high tunnel and open field) were significant at $p \leq 0.01$, with a linear least squares fit equation of 1.66X + 0.77 and $R^2 = 0.26$. The data show a weak correlation between the TA and acidity (PAM) data (Figure 2A); however, when the outliers were removed (Figure 2B), the R^2 increased to 0.47, with the equation of 2.22X + 0.49, where X is the acidity.

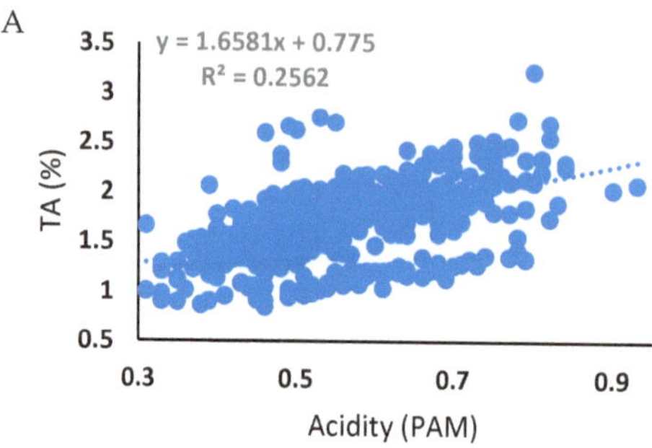
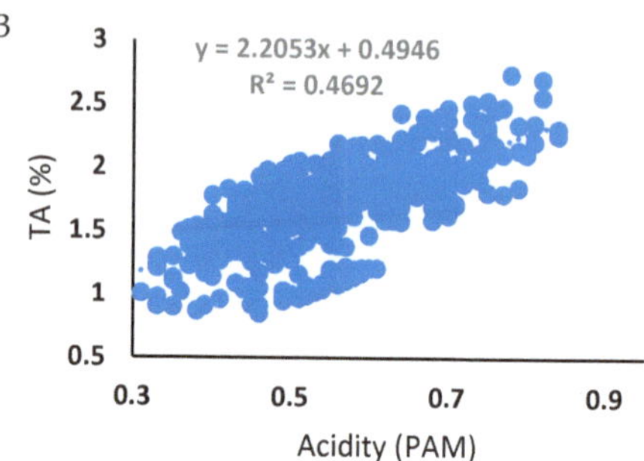

Figure 2. The regression between acidity measured using the PAM and the titratable acidity (TA) measured using titration with NaOH 0.1 N. (**A**) All data were included in the graph; (**B**) outliers were removed. The R^2 increased to 0.4692 when the outliers were removed.

The regression data for each cultivar in each environment showed that, in high tunnels, cultivars were different, either not showing any correlation between acidity (PAM) and TA, such as for "Chandler" (R^2 = 0.0004), or showing a medium level of correlation, such as for "Rocco" (R^2 = 0.32). However, in the open field, all cultivars except "Rocco" showed more correlation than the high tunnel and R^2, ranged between 0.20 and 0.38 (Figure 3). For example, "Chandler" showed a much higher correlation (R^2 = 0.31) between acidity (PAM) and TA content in the open field compared with the high tunnel. The cultivar "Ruby June" had the highest correlation between TA and acidity (PAM) in the open field, with a linear least squares fit equation of 2.35X + 0.49 and an R^2 of 0.38.

Our results indicated that the pocket meter can be a very rapid means of estimating titratable acidity in strawberries. The results confirmed the previous report by Perkins-Veazie et al. [17] that the correlations between TA and acidity (PAM data) were not strong for strawberries and there were some inconsistencies. Our results show that acidity data were more consistent for strawberries grown in the open field than for those grown in the high-tunnel environment. There are no solid results on cultivars, with some showing a low correlation in the high tunnel and a higher correlation in the open field. The influence of harvest date or organic acid profile of the strawberry germplasm may be essential for further studies.

Ideally, the PAM data should match the titrimeter values. This was not always the case for these data and we could not predict what caused the outliers within each cultivar of strawberries or the environment. The PAM data had only two decimal places, whereas the automated titrimeters yielded four. If very precise values for TA are needed, such as when developing value-added products, then an automated titrimeter would be the better choice. An added attraction is that two or three of the PAM meters could be used by several people at minimal cost and further accelerate the collection of titratable acidity data. Further, using a PAM does not require special chemicals or a specialized laboratory.

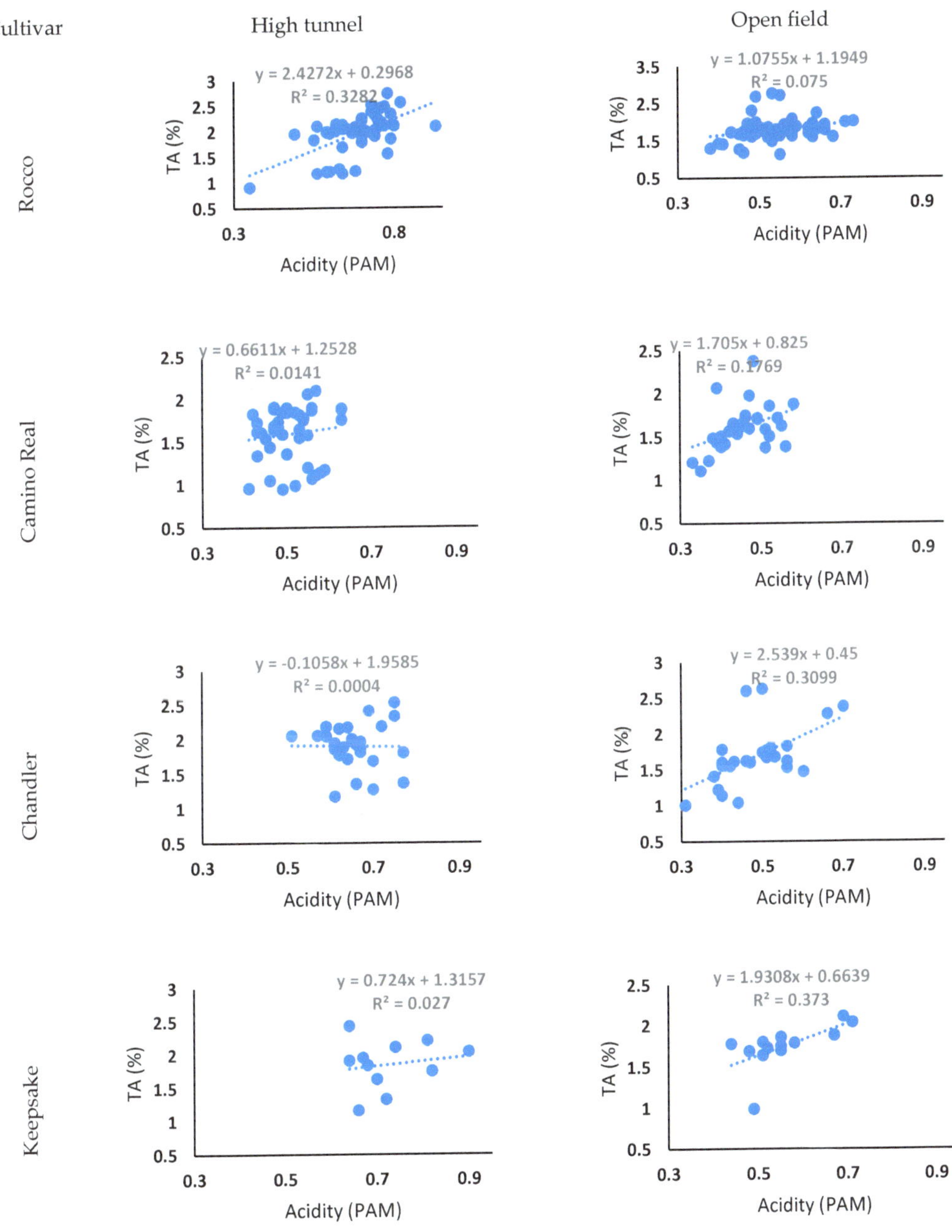

Figure 3. *Cont.*

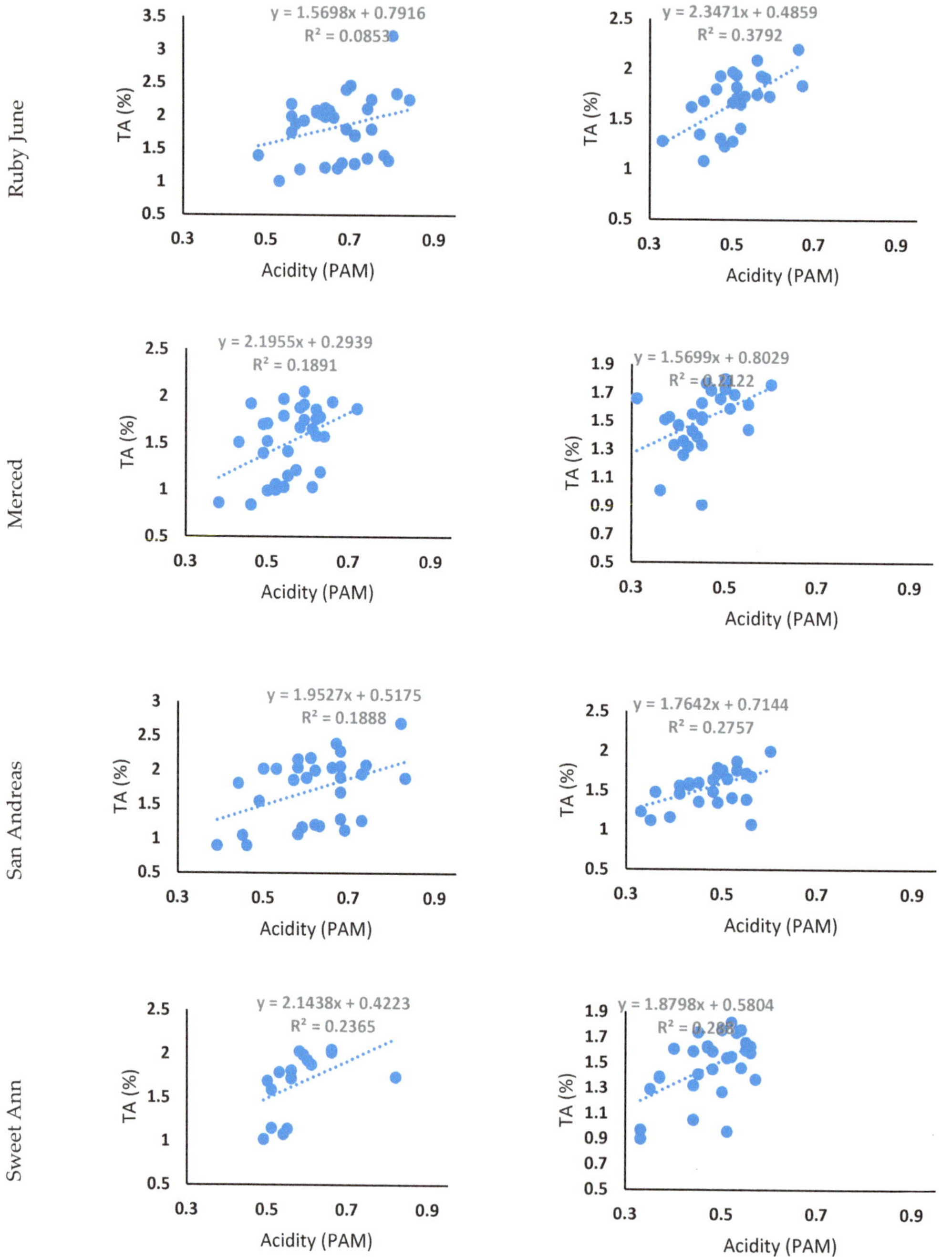

Figure 3. The regression between acidity (PAM data) and the titratable acidity (TA) measured by titration with NaOH 0.1 N for cultivar and environment interactions. The cultivars "Flavorfest" and "Albion" did not have enough representations in the samples.

4. Conclusions

Genetics substantially affects strawberry fruit quality (total soluble solids, TA, and acidity) more than the two environments. Strawberry cultivars had higher anthocyanin content and acidity in the high tunnel than in the open field. The environment (high tunnel and open field) did not have a significant effect on TSS. Anthocyanin and TSS contents were higher in "Flavorfest" than in other cultivars. "Sweet Ann" and "San Andreas" had the lowest TSS and TA. The fruit quality parameters did not significantly change during the harvest season.

Although the correlation between the TA and acidity (PAM data) was not strong, the TA can be estimated using PAM efficiently and with little effort. The estimate will fulfill the industry requirements for identifying the harvest date (especially in the open field) considering that the acidity level is not a strong indicator of fruit quality. The reason for the weak correlation is not understood and warrants further investigation.

Author Contributions: Conceptualization, methodology, software, validation, resources, data curation, visualization, supervision; and funding acquisition, T.T. and J.B.S.; formal analysis, T.T.; investigation, H.P., T.T. and J.B.S.; writing-first draft T.T.; supervision, T.T. All authors have read and agreed to the published version of the manuscript.

Funding: This research was funded by the USDA Specialty Crop Block Grant, USDA-NIFA-Capacity Building Grant (grant number 2019-38821-29038) and the USDA-NIFA-Evans Allen (grant numbers 1015372). Virginia State University provided staff salary and support.

Data Availability Statement: All available data are presented in the manuscript.

Acknowledgments: The authors thank Danyang Liu, Aman Rana, Sophia Gonzales, Robert Holtz, and Greyson Dockiewicz for their assistance in field plot maintenance. This article is a contribution of the Virginia State University (VSU) Agricultural Research Station (Journal Article Series Number 394).

Conflicts of Interest: The authors declare no conflict of interest.

References

1. Shahbandeh, M. Leading Strawberry Producing U.S. States, Strawberry Production in the United States in 2020 by State, 2020. Available online: https://www.statista.com/statistics/194235/top-10-strawbery-producing-us-states/ (accessed on 18 May 2021).
2. Food and Agriculture Organization of the United Nations. 2015. Crops. Available online: http://www.fao.org/faostat/en/#data/QC (accessed on 28 November 2018).
3. U.S. Department of Agriculture (USDA). World Strawberry Production, 1990–2011. 2013. Available online: http://usda.mannlib.cornell.edu/MannUsda/viewDocumentInfo.do?documentID=1381 (accessed on 28 November 2018).
4. Samtani, J.B.; Rom, C.R.; Friedrich, H.; Fennimore, S.A.; Finn, C.E.; Petran, A.; Wallace, R.W.; Pritts, M.P.; Fernandez, G.; Chase, C.A.; et al. The status and future of the strawberry industry in the United States. *HortTechnology* **2019**, *29*, 11–24. [CrossRef]
5. Flanagan, R.D.; Samtani, J.B.; Manchester, M.A.; Romelczyk, S.; Johnson, C.S.; Lawrence, W.; Pattison, J. On-farm evaluation of strawberry cultivars in coastal Virginia. *HortTechnology* **2020**, *30*, 789–796. [CrossRef]
6. Larson, K.D. Strawberry. In *Handbook of Environmental Physiology of Fruit Crops*; CRC Press: Boca Raton, FL, USA, 2018; pp. 271–297.
7. Cervantes, L.; Ariza, M.T.; Miranda, L.; Lozano, D.; Medina, J.J.; Soria, C.; Martínez-Ferri, E. Stability of fruit quality traits of different strawberry varieties under variable environmental conditions. *Agronomy* **2020**, *10*, 1242. [CrossRef]
8. Paroussi, G.; Voyiatzis, D.G.; Paroussis, E.; Drogoudi, P.D. Growth, flowering and yield responses to GA3 of strawberry grown under different environmental conditions. *Sci. Horticult.* **2002**, *96*, 103–113. [CrossRef]
9. Taghavi, T.; Patel, H.; Rafie, R. Comparing pH differential and methanol-based methods for anthocyanin assessments of strawberries. *Food Sci. Nutr.* **2022**, *10*, 2123–2131. [CrossRef] [PubMed]
10. Voča, S.; Dobričević, N.; Dragović-Uzelac, V.; Duralija, B.; Družić, J.; Čmelik, Z.; Babojelić, M.S. Fruit Quality of New Early Ripening Strawberry Cultivars in Croatia. *Food Technol. Biotechnol.* **2008**, *46*, 292–298.
11. Oregon Strawberry Commission. Product Development Guide, 2006; pp. 1–4. Available online: www.oregon-strawberries.com (accessed on 15 June 2022).
12. Keutgen, A.J.; Pawelzik, E. Modifications of taste-relevant compounds in strawberry fruit under NaCl salinity. *Food Chem.* **2007**, *105*, 1487–1494. [CrossRef]
13. Gündüz, K.; Özbay, H. The effects of genotype and altitude of the growing location on physical, chemical, and phytochemical properties of strawberry. *Turk. J. Agric. For.* **2018**, *42*, 145–153. [CrossRef]
14. Pelayo-Zaldívar, C.L.A.R.A.; Ebeler, S.E.; Kader, A.A. Cultivar and harvest date effects on flavor and other quality attributes of California strawberries. *J. Food Qual.* **2005**, *28*, 78–97. [CrossRef]

15. Chaiwong, S.; Bishop, C.F. Effect of vibration damage on the storage quality of 'Elsanta' strawberry. *Aust. J. Crop Sci.* **2015**, *9*, 859–864.

16. Perkins-Veazie, P.; Ashrafi, H.; Fernandez, G.; Clark, J.R.; Threfall, R.; Worthington, M.; Taghavi, T. Multi User Testing of a Pocket Acidity Refractometer (PAM) as a Rapid Means to Determine Titratable Acidity in Small Fruits, Final Report for Southern Region Small Fruit Consortium Grant, 2019. Available online: https://smallfruits.org/files/2019/12/2019-SRFRC-titrmeter-study.pdf (accessed on 19 April 2022).

17. Taghavi, T.; Patel, H.; Akande, O.E.; Galam, D.C.A. Total Anthocyanin Content of Strawberry and the Profile Changes by Extraction Methods and Sample Processing. *Foods* **2022**, *11*, 1072. [CrossRef] [PubMed]

18. SAS Institute Inc. *SAS/ACCESS®, 9.4*; SAS Institute Inc.: Cary, NC, USA, 2013.

19. Singh, A.; Syndor, A.; Deka, B.C.; Singh, R.K.; Patel, R.K. The effect of microclimate inside low tunnels on off-season production of strawberry (*Fragaria* × *ananassa* Duch.). *Sci. Hortic.* **2012**, *144*, 36–41. [CrossRef]

20. Hanson, E.; Von Weihe, M.; Schilder, A.C.; Chanon, A.M.; Scheerens, J.C. High tunnel and open field production of floricane-and primocane-fruiting raspberry cultivars. *HortTechnology* **2011**, *21*, 412–418. [CrossRef]

21. Zhang, Y.; Yang, M.; Hou, G.; Zhang, Y.; Chen, Q.; Lin, Y.; Li, M.; Wang, Y.; He, W.; Wang, X.; et al. Effect of Genotype and Harvest Date on Fruit Quality, Bioactive Compounds, and Antioxidant Capacity of Strawberry. *Horticulturae* **2022**, *8*, 348. [CrossRef]

22. Herrington, M.E.; Chandler, C.K.; Moisander, J.A.; Reid, E.C. Rubygem strawberry. *Hortscience* **2007**, *42*, 1482–1483. [CrossRef]

23. Saraçoğlu, O. Determination of Yield and Quality Performance of Some Neutral and Short Day Strawberry Cultivars in Kazova. Ph.D. Thesis, University of Gaziosmanpaşa, Tokat, Turkey, 2013.

24. Andreotti, C.; Guerrero, G.; Zago, M. Quality of strawberry fruits cultivated in a highland area in South Tyrol (Italy). *Acta Hort.* **2014**, *1049*, 795–799. [CrossRef]

25. Azodanlou, R. A Methodology for Assessing the Quality of Fruit and Vegetables. Ph.D. Thesis, Swiss Federal Institute of Technology, Zurich, Switzerland, 2011. Available online: https://www.research-collection.ethz.ch/bitstream/handle/20.500.118 50/145336/eth-24231-02.pdf (accessed on 30 July 2023).

26. Kallio, H.; Hakala, M.; Pelkkikangas, A.-M.; Lapveteläinen, A. Sugars and acids of strawberry varieties. *Eur. Food Res. Technol.* **2000**, *212*, 81–85. [CrossRef]

Article

Influence of Sunn Hemp Biomass Incorporation on Organic Strawberry Production

Yurui Xie, Zachary E. Black, Nan Xu, Jeffrey K. Brecht, Dustin M. Huff and Xin Zhao *

Horticultural Sciences Department, University of Florida, Gainesville, FL 32611, USA
* Correspondence: zxin@ufl.edu

Abstract: Sunn hemp (*Crotalaria juncea* L.), a warm season leguminous cover crop, is commonly used in rotation with organic strawberry production in Florida's subtropical environment. This study was conducted to explore the impacts of sunn hemp on growth and yield performance of the subsequent organic strawberry crop in sandy soils, taking into consideration the nutrient contribution from soil incorporation of sunn hemp biomass. Sunn hemp was seeded during the summer off-season and terminated before flowering, three weeks prior to the fall planting of two strawberry cultivars ('Strawberry Festival' and 'Camino Real'). With sunn hemp residues incorporated into the soil, two application rates of nitrogen (N) through pre-plant organic fertilization for the strawberry season were used, including N at a rate of 84 kg/ha, without consideration of the N credit from sunn hemp, and N at a rate of 19.8 kg/ha, with consideration of the estimated N credit from sunn hemp. A summer fallow without cover crop and with a pre-plant organic fertilizer application at the N rate of 84 kg/ha was included as the control. Overall, the sunn hemp incorporation at three weeks after termination did not benefit the strawberry plant growth or fruit yield in this study, with rather low levels of soilborne pathogen and nematode infestations. Both sunn hemp treatments exhibited a significantly lower level of total soil N compared to the summer fallow plots at the end of the strawberry season. The reduction in the pre-plant N fertilization resulted in lower above-ground plant dry weight and accumulation of N, phosphorus (P), and potassium (K) at the end of the strawberry season, along with fewer leaves and smaller crowns of the strawberry plants during the early season. Both sunn hemp treatments decreased early fruit yields, while the sunn hemp treatment with the reduced N fertilization also led to a significant reduction in the total fruit number and weight, although no significant differences in the whole-season marketable fruit yield were observed among the nutrient management treatments. Overall, 'Strawberry Festival' yielded higher than 'Camino Real', but the effects of nutrient management did not vary with the strawberry cultivars. Further studies are needed to enhance organic strawberry nutrient management involving rotational cover crops.

Keywords: *Fragaria* ×*ananassa* Duch.; *Crotalaria juncea* L.; cultivar; fruit yield; growth; nutrient management; pre-plant fertilization

Citation: Xie, Y.; Black, Z.E.; Xu, N.; Brecht, J.K.; Huff, D.M.; Zhao, X. Influence of Sunn Hemp Biomass Incorporation on Organic Strawberry Production. *Horticulturae* **2023**, *9*, 1247. https://doi.org/10.3390/horticulturae9111247

Academic Editor: Toktam Taghavi

Received: 28 September 2023
Revised: 24 October 2023
Accepted: 31 October 2023
Published: 20 November 2023

1. Introduction

Strawberry (*Fragaria* ×*ananassa* Duch.), one of the most valuable small fruit crops in the United States, is also a top organic commodity, reaching 2145 ha in harvested acreage and $336 million in value of sales in 2021 [1]. Florida is a leading state in winter strawberry production focused on the high-value early market. Growers' interest in organic strawberry production has been increasing in Florida in recent years, as shown by the increase in the number of certified organic farms according to the recent U. S. organic production surveys [1].

Cover cropping is an essential part of soil quality and fertility management practices in organic crop production systems. The appropriate use of cover crops improves soil organic matter (OM) and health, reduces soil compaction and erosion, and suppresses weeds. Leguminous cover crops are also employed as green manure, contributing to increased

N availability [2,3]. The integration of cover crop residues into nutrient management programs for organic crop production promotes nutrient cycling and helps optimize the benefits of using rotational cover crops. An estimation of 90–100 kg/ha of N could be supplied annually by the winter legume cover crop hairy vetch (*Vicia villosa* Roth.) to no-tillage corn [4]. Different legumes and grass species not only differ in their biomass accumulation, but also vary in their N release rate [5]. In addition to environmental factors such as soil temperature and moisture, the rate of N mineralization and release from cover crop residues is largely determined by the carbon to nitrogen (C:N) ratio of the cover crop at termination. A C:N ratio near 24:1 facilitates microbial digestion to achieve a relatively fast breakdown of plant residues, whereas a larger C:N may result in a temporary N deficit to the following cash crops due to N immobilization [6,7]. On the other hand, a C:N ratio lower than 24:1 may speed up N release early in the growing season, making it difficult to meet crop N demand during yield development.

Growing cover crops prior to the strawberry season has been shown to be an effective tool for weed management; but previous studies have reported different results regarding the cover crop effects on the growth and yield of strawberry plants. Sudangrass (*Sorghum bicolor* L.), pearl millet (*Pennisetum glaucum* L.), soybean (*Glycine max* L.), or velvetbean [*Mucuna deeringiana* (Bort) Merr.] utilized as annual summer rotational cover crops were found to suppress summer weed populations but had no effect on organic strawberry growth or yield in North Carolina when short-day 'Chandler' strawberry plants were transplanted into plasticulture beds in the fall [8]. In contrast, a study conducted in Iowa indicated that sudangrass, big bluestem (*Andropogon gerardii* Vitman), or switchgrass (*Panicum virgatum* L.) that had been grown for multiple years with a final incorporation into the soil the summer before strawberry planting not only reduced fall weed population and biomass, but also improved plant establishment and increased fruit yield of the conventional June-bearing 'Honeoye' strawberry crop in a matted-row system [9].

In Florida, summer cover crops have been used in rotation with the fall planting of strawberries by organic growers, with sunn hemp (*Crotalaria juncea* L.) being the most common one due to its suppressive effects on sting nematodes (*Belonolaimus longicaudatus* Rau.), a major pest in Florida strawberry production [10]. Sunn hemp residues exhibited allelopathic effects on the seed germination of weeds and certain vegetable crops, but there are no known effects on field-grown strawberries. Sunn hemp can produce 5050–11,235 kg/ha of dry biomass with about 2.85% of N in plant residues and provide 100–200 kg/ha of total available N to the subsequent cash crop [11–13]. Sunn hemp was considered to be a well-suited cover crop in the Southeastern U.S. which could provide 33% to 50% of the total N needed for most crops, assuming a 50% availability of the N accumulated from sunn hemp after 60 days of seeding [13]. However, despite the popularity of using sunn hemp in rotation with organic strawberry cultivation, research-based information is scarce regarding its effects on the plant performance of the following strawberry crop, as well as on the role of sunn hemp in the nutrient management of organic strawberry plants given its considerable contribution of N release.

In this study, the effects of sunn hemp as a summer rotational crop on the subsequent strawberry production were assessed by considering the nutrient availability from sunn hemp biomass accumulation in terms of plant growth and fruit yield of two strawberry cultivars in an organic production system.

2. Materials and Methods

2.1. Experimental Design and Field Trial Establishment

The field trial was conducted during the 2014–2015 strawberry season on a certified organic land at the University of Florida Plant Science Research and Education Unit (PSREU) in Citra, FL, USA (lat. 29.41° N, long. 82.16° W). This region has a humid subtropical climate, and the Candler sand soil at the experimental site is classified as arenosols according to the World Reference Base for Soil Resources (WRB). During the experimental period, the average daily air temperature was 19.3 °C and the accumulative precipitation reached 745.5 mm (Figure 1).

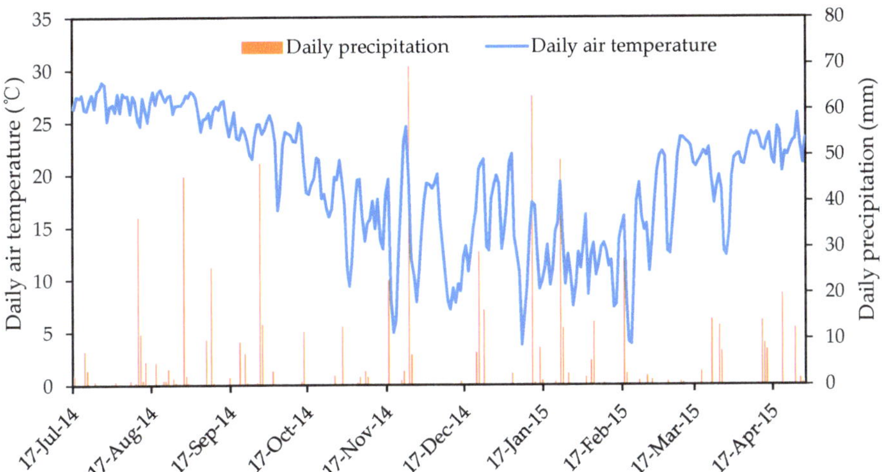

Figure 1. Daily air temperature and precipitation in Citra during the entire experimental period. The data reported were from 17 July 2014 to 29 April 2015. Data source: Florida Automated Weather Network (FAWN): https://fawn.ifas.ufl.edu/ (accessed on 31 October 2023).

The field experiment was arranged in a split-plot design with four replications and eighty plants per subplot. The soil and nutrient management practice for organic strawberry production was the whole plot factor, while the strawberry cultivar was the subplot factor. There were three soil and nutrient management treatments in the whole plots, which were randomized in a complete block design, including: (1) sunn hemp as a summer cover crop before the strawberry production season, with a pre-plant N fertilization rate at 84.0 kg N/ha for the strawberry season; (2) sunn hemp as a summer cover crop before the strawberry production season, with a reduced pre-plant N fertilization providing 19.8 kg/ha of N, assuming that 50% of the N from sunn hemp incorporation would be available for strawberry uptake, based on the sunn hemp biomass estimation and tissue analysis for nutrients; and (3) a summer fallow control without sunn hemp and with a pre-plant fertilization rate at 84.0 kg/ha of N. The two short-day strawberry cultivars used in this study were 'Strawberry Festival' and 'Camino Real'.

The sunn hemp was broadcast seeded at 44.8 kg/ha on 17 July 2014, and the plants were incorporated into the soil using flail-mowing and roto-tilling at a depth of about 13 cm on 22 September 2014, at which time < 5% of the plants started to flower. The plots were tilled twice more at a 13 cm depth before the bed formation and the strawberry transplanting on 14 October 2014. The plots without sunn hemp were maintained as a summer fallow before strawberry planting. In order to minimize the weed pressure effect, the summer fallow plots were tilled three times at a soil depth of approximately 13 cm on 13 August, 7 September, and 14 October 2014. All the field plots were hand-weeded during the strawberry season.

The organic fertilizers used for the pre-plant application included a mixture of MicroSTART60 3N-0.9P-2.5K (Perdue AgriRecycle, LLC., Seaford, DE, USA), Howard Organic Bonemeal 7N-5.2P-0K (Howard Fertilizer & Chemical Co., Inc., Orlando, FL, USA), and Jobe's Organics Bone Meal 2N-6.1P-0K (Easy Gardener Products, Inc., Waco, TX, USA). They were banded in the bed location and incorporated during bed formation. The pre-plant phosphorus (P) and potassium (K) fertilization rates were determined according to the soil test results. For the reduced pre-plant fertilization treatment, the P and K availability from the sunn hemp residues was also factored into the calculation. GATOR 96002 Organic

Liquid 3N-0P-5.0K (Howard Fertilizer & Chemical Co., Inc., Orlando, FL, USA) was used for in-season fertigation through drip irrigation. The in-season fertilization for the organic strawberry production remained the same for all treatments. All the fertilizer products utilized are approved for use in certified organic crop production. The pre-plant application rates of N, P, and K, as well as the application rates throughout the whole strawberry season for each soil and nutrient management treatment, are presented in Table 1.

Table 1. Nitrogen (N), phosphorus (P), and potassium (K) sources and rates for full and reduced pre-plant fertilization treatments used in the organic strawberry field trial in Citra, FL, USA.

	N (kg/ha)	P (kg/ha)	K (kg/ha)	Organic Fertilizer and Ratio (Based on Weight)
Pre-plant fertilization [1]				
Full N rate	84.0	45.5	21.4	MicroSTART60: Howard Organic Bone Meal = 3:2
Reduced N rate	19.8	38.8	5.6	Jobe's Organics Bone Meal: MicroSTART60 = 2:1
In-season fertigation [2]	152.2	0.0	253.6	GATOR 96002 Organic Liquid 3N-0P-5.0K
Whole-season fertilization [3]				
Full N rate	236.2	45.5	275.0	Pre-plant fertilization at the full rate plus fertigation
Reduced N rate	172.0	38.8	259.2	Pre-plant fertilization at the reduced rate plus fertigation

[1] Part of nutrient management treatments, applied before strawberry transplanting on 14 October 2014; [2] Liquid fertilizer applied after the strawberry plant establishment through drip irrigation. The same fertigation rates were used in both the full N rate and reduced N rate treatments; [3] Total fertilization applied during the whole strawberry season, including the pre-plant and in-season fertilization.

2.2. Field Planting of Strawberry

There were four 26.7 m long planting beds in each replication (block), with each planting bed being divided into three sections for the three soil and nutrient management treatments; two nearby sections of two parallel beds represented a subplot (80 strawberry plants), where each of the strawberry cultivars was planted. Strawberry plug plants (Luc Lareault Nursery, Lavaltrie, QC, Canada) were transplanted into double rows spaced 30.5 cm apart on the raised beds on 14 October 2014. The planting beds were 1.0 m wide at the base and 0.8 m wide at the top, 17.8 cm high, and spaced 1.5 m apart from their center The beds were covered with 1.25 mil black polyethylene mulch (Intergro, Inc., Clearwater, FL, USA). Timer-controlled irrigation was applied twice per day, for 45 min (min) each, and the irrigation schedule was adjusted as needed. The plants were fertigated through a drip irrigation system (30.5 cm emitter spacing) under the plastic mulch at the N application rate of 0.67 kg/ha per day starting on 14 November 2014, later increased to 1.12 kg/ha per day from 5 December 2014, and finally adjusted to 1.34 kg/ha per day from 6 March 2015 until the end of the season. AgroFabric Pro42 row covers that transmit 60% of light (Universal Enterprises Supply, Pompano Beach, FL, USA) were applied for frost protection. The predatory mites *Phytoseiulus persimilis* and *Neoseiulus californicus* (Spidex and Spical; Koppert Biological Systems, Inc., Howell, MI, USA) were released on 21 November 2014 to control the two-spotted spider mites.

2.3. Soil and Plant Tissue Analyses

The sunn hemp was sampled for above- and below-ground tissue in five randomly selected areas from each of the whole plots with sunn hemp being grown in the summer, using 0.5 m × 0.5 m quadrants. After drying the fresh tissue at 65 °C for two weeks to a constant weight, the total amount of dry weight of the sunn hemp was recorded. The dried sunn hemp samples were sent to Waters Agricultural Laboratories, Inc. (Camilla, GA, USA) for measuring the N, P, and K contents. The total N was analyzed using the dry-combustion method, while the P and K contents were analyzed with the open-vessel wet-digestion method, followed by an inductively coupled argon plasma (ICAP) spectroscopy analysis. The total amounts of N, P, and K provided by the sunn hemp were calculated based on the dry biomass and corresponding tissue nutrient analysis results.

Soil samples were collected at a 30 cm depth for soil nutrient analysis twice during the season, as follows: before sunn hemp incorporation (22 September 2014) and after the final harvest (29 April 2015). The first soil sampling comprised three replicated samples from each of the twelve whole plots, and the second soil sampling was composed of three replicated samples from each of the twenty-four subplots. The soil organic matter (OM) content, cation exchange capacity (CEC), total soil N, and levels of extractable P, K, calcium (Ca), magnesium (Mg), sulfur (S), copper (Cu), iron (Fe), manganese (Mn), zinc (Zn), and boron (B) were analyzed using the Mehlich-1 extraction method (Waters Agricultural Laboratories, Inc., Camilla, GA, USA). The soil test results before strawberry planting were used to develop the soil fertility management program for the strawberry season.

Strawberry tissue macronutrient (N, P, K, Ca, Mg, and S) and micronutrient (Fe, Mn, Zn, Cu, and B) analyses were conducted (Waters Agricultural Laboratories, Inc.) in the early season (26 November 2014), 43 days after transplanting (DAT) and after the final harvest (22 April 2015, 190 DAT) by sampling 14 of the most recently matured leaves from each subplot. The nutrient contents of the dried strawberry leaf samples were determined following the same protocols used for the sunn hemp tissue analyses.

2.4. Strawberry Growth and Yield Assessments

The strawberry plants showing open flowers were counted on 4, 10, and 17 November and on 2 December 2014. Four randomly selected strawberry plants from each subplot were marked on 14 November 2014 for growth assessments including leaf number, canopy size, crown diameter, and chlorophyll content index at different stages, as follows: early season (14 November and 4 December 2014, 31 and 51 DAT, respectively), early and peak harvest (18 December 2014 and 22 January 2015, 65 and 100 DAT, respectively), and late season (12 March and 22 April 2015, 149 and 190 DAT, respectively). Only the fully matured leaves that emerged after transplanting were counted. The smallest and the largest canopy diameters were measured, and the average of the two was used to determine the canopy size. The overall crown diameter was measured at approximately 1 cm above the soil line using an electronic caliper (General Tools & Instruments, Secaucus, NJ, USA). The leaf chlorophyll content index was were examined on ten most recently matured leaves in each subplot using a SPAD 502 Plus Chlorophyll Meter (Spectrum Technologies, Inc., Aurora, IL, USA). The above- and below-ground biomass of the strawberry plants, excluding flowers or fruit, was assessed on 1 May 2015 after the final harvest. Three randomly selected plants from each subplot were collected from the field, cleaned, separated at the soil line, and oven-dried at 60 °C for four weeks to a constant weight. The dry samples were then weighed to estimate the plant biomass accumulation over the whole season. The above-ground plant accumulation of N, P, and K was also estimated based on the dry biomass and leaf tissue nutrient contents.

Fruit with calyx attached were picked from all the field plots. The strawberries were harvested from 4 December 2014 to 6 April 2015, approximately twice per week, with a total of 26 harvests conducted. The fruit number and weight were measured to determine marketable and unmarketable fruit yields. The strawberry fruit (>5 g) with red color over at least 80% of the surface area and without decay, disease, pest, or mechanical damages was considered as marketable fruit.

2.5. Statistical Analyses

The data analysis was conducted using the Glimmix procedure of the SAS statistical software package for Windows (Version 9.2; SAS Institute, Cary, NC, USA). A two-way analysis of variance (ANOVA) was performed following the split-plot design. The Fisher's Least Significant Difference (LSD) test was used for multiple comparisons of different measurements among treatments at $\alpha = 0.05$.

3. Results and Discussion

3.1. Soil and Plant Tissue Analyses

The soil tests before sunn hemp incorporation showed comparable levels of soil OM, CEC, total N, and extractable P, Ca, Mg, S, Fe, Mn, Zn, Cu, and B between the sunn hemp plots and the summer fallow plots, but the former had a significantly lower level of available K in the soil (Table 2). A previous study conducted in Florida sandy soils also reported a significant reduction in the soil test of K levels following cover cropping over the summer season [14]. After the final strawberry harvest, i.e., about seven months following the soil incorporation of the sunn hemp residues, similar levels of soil OM, CEC, and available nutrients were observed among the sunn hemp treatments and the summer fallow control, except for the total soil N level. While the total soil N level was similar among the sunn hemp plots despite their different pre-plant fertilization rates, it was significantly lower compared to the summer fallow plot. This finding was surprising given the total amount of above-ground dry biomass (4620.5 kg/ha) of sunn hemp and the N accumulation (117.1 kg/ha) at soil incorporation prior to the strawberry planting. The above-ground accumulation of N in sunn hemp observed in the present study is within the commonly reported range of 73–207 kg/ha [15]. The average C:N ratio of the sunn hemp residues at termination was 21:1, suggesting that a relatively fast decomposition might have occurred after soil incorporation [11–13,16]. According to Wang et al. (2011) [17], the breakdown of sunn hemp residues usually takes place within two weeks following soil incorporation in tropical areas. A previous study in Florida sandy soils also showed that approximately 50% of N could be released during the first four weeks after sunn hemp residue incorporation [18]. Another field decomposition study of sunn hemp residues demonstrated that as much as 64% of N could be released within the first two weeks following sunn hemp termination, with the total N release reaching 79% after 6 months [19]. Moreover, N mineralization and release from organic amendments is strongly correlated with soil temperature and moisture, being generally faster in warmer and moist soils until the soil water potential reaches its maximum [20–24]. In this study, the daily average soil temperature at 10-cm depth ranged from 13.5 to 28.3 °C, and 107.2 mm of rainfall occurred between sunn hemp termination (22 September 2014) and four weeks after strawberry transplanting (11 November 2014) [25]. Given that the active plant uptake of nutrients, particularly N, may not occur until the field establishment of the strawberry crop (at least 7–10 DAT), it was likely that most of the mineralized N from the sunn hemp residues had already been lost from the root zone through denitrification and leaching prior to strawberry transplanting. It was reported that the fast release of available N from sunn hemp was likely exceeding the demand of strawberry plants and could potentially result in N losses [26].

Table 2. The soil organic matter (OM), cation exchange capacity (CEC), and nutrient levels before sunn hemp incorporation (22 September 2014) and after the strawberry final harvest (29 April 2015) in Citra, FL, USA.

Treatment	OM (%)	CEC (meq/100 g)	TN (%)	P	K	Ca	Mg	S	Fe	Mn	Zn	Cu	B
					Before Sunn Hemp Incorporation			(mg/kg)					
Management (M)													
SF w/full N	0.77	3.8	0.04	35.9	10.4 a	453.6	56.1	17.4	8.0	2.1	0.3	0.1	0.1
SH w/full N	0.76	3.7	0.05	35.1	8.8 b	460.8	52.9	11.0	9.5	1.9	0.3	0.1	0.1
SH w/reduced N	0.79	3.8	0.07	35.9	8.8 b	471.4	53.5	11.6	8.3	1.7	0.3	0.1	0.1
Significance	NS	NS	NS	NS	*	NS	NS	NS	NS	NS	NS	NS	NS

Table 2. *Cont.*

Treatment	OM (%)	CEC (meq/100 g)	TN (%)	After Strawberry Final Harvest									
				P	K	Ca	Mg	S (mg/kg)	Fe	Mn	Zn	Cu	B
Management (M).													
SF w/full N	0.67	4.5	0.19 a	50.6	76.5	588.5	31.3	14.2	7.3	2.0	0.8	0.2	0.2
SH w/full N	0.78	4.9	0.15 b	45.9	94.1	661.3	32.4	6.7	6.9	1.9	0.8	0.2	0.2
SH w/reduced N	0.74	4.9	0.15 b	49.2	89.3	664.1	33.8	7.9	7.0	2.3	1.0	0.2	0.2
Significance	NS	NS	*	NS	NS	NS	NS	NS	NS	NS	NS	NS	NS
Cultivar (C)													
Strawberry Festival	0.73	4.7	0.16	47.8	80.3 b	622.8	32.1	10.9	7.1	2.1	0.8	0.2	0.2
Camino Real	0.73	4.8	0.16	49.3	93.0 a	653.1	32.8	8.3	7.0	2.0	0.9	0.2	0.2
Significance	NS	NS	NS	NS	**	NS	NS	NS	NS	NS	NS	NS	NS
M × C interaction	NS	NS	NS	NS	NS	NS	NS	NS	NS	NS	NS	NS	NS

SF: summer fallow; SH: sunn hemp; OM: organic matter; CEC: cation exchange capacity; TN: total nitrogen. The means within the same column followed by the same letter do not differ significantly according to Fisher's least significant difference test at $p \leq 0.05$. NS, *, and **: nonsignificant or significant at $p \leq 0.05$ or 0.01, respectively.

With regard to the strawberry cultivar impact on the soil nutrients after the final harvest, 'Camino Real' resulted in a significantly higher level of soil K than 'Strawberry Festival' (Table 2), suggesting a possibly higher demand of K by 'Strawberry Festival'. The sting nematode infestation did not interfere with the treatments in this study as the soil tests of each subplot at the end of the strawberry season showed an undetectable level of sting nematode population. The soil pH did not differ significantly among the different soil nutrient management treatment plots or between the different cultivar plots; however, it showed an increase, from 6.4 to 7.4, from before the sunn hemp incorporation to after the strawberry harvest, during the experimental period. The strawberry leaf tissue nutrient contents were not significantly affected in the early season (43 DAT) and after the final harvest (190 DAT) by different soil and nutrient management practices, except for the Mn and S after the final harvest. The leaf Mn levels were significantly higher in the sunn hemp treatment with the reduced pre-plant N fertilization compared to the treatment with the full N fertilizer rate, as well as the summer fallow plots. The strawberry leaf content of S was significantly higher in the summer fallow plots relative to the sunn hemp plots (Table 3). In contrast, strawberry cultivar had a greater impact on leaf nutrient contents, although the varietal difference varied during the production season (Table 3). The early season leaf tissue analysis showed that the strawberry plants were adequate in the levels of N, K, Ca, S, B, Mn, and Fe, high in P and Mg, but slightly deficient in Zn and Cu [27]. The leaf tissue contents of P, Cu, and Zn were significantly higher in 'Camino Real' than 'Strawberry Festival' in the early season, whereas 'Strawberry Festival' had higher levels of leaf P, Mg, S, Fe, Mn, and B after the final harvest (Table 3). The soil and nutrient management by cultivar interaction was detected for Zn in the early season, as shown by the higher level of Zn in the summer fallow control for 'Camino Real' but not 'Strawberry Festival'. The interaction effects were also found in the strawberry leaf P, S, and B contents after the final harvest. The 'Strawberry Festival' grown with the reduced pre-plant N fertilization had significantly lower levels of P and S compared to the full pre-plant N fertilization treatments with or without sunn hemp, while the sunn hemp with the full pre-plant N fertilization resulted in the highest B content in the 'Strawberry Festival' leaves but the lowest B level in 'Camino Real'. The total N, P and K accumulation in the strawberry leaf tissue after the final harvest showed a significant reduction in N, P, and K in the plants grown in the reduced pre-plant N fertilization plots in contrast to the full N rate plots. No difference was observed between the two strawberry cultivars (Table 4).

Table 3. Nutrient management and cultivar effects on nutrient contents of the most recently matured leaves in the early season (26 November 2014, 43 DAT) and after the final harvest (22 April 2015, 190 DAT) in the organic strawberry field trial in Citra, FL, USA.

	N	P	K	Ca	Mg	S	Fe	Mn	Zn	Cu	B
						Early Season					
Treatment			(g/100 g)						(mg/kg)		
Management (M)											
SF w/full N	3.18	0.47	1.75	1.41	0.57	0.21	73.6	45.3	16.3	3.4	30.3
SH w/full N	3.20	0.44	1.85	1.41	0.58	0.21	72.6	46.0	15.0	3.6	31.4
SH w/reduced N	3.15	0.44	1.83	1.46	0.59	0.20	69.4	39.0	14.4	3.5	30.9
Significance	NS	NS	NS	NS	NS	NS	NS	NS	NS	NS	NS
Cultivar (C)											
Strawberry Festival	3.21	0.41 b	1.85	1.49	0.60	0.21	70.6	42.7	13.7 b	3.3 b	30.8
Camino Real	3.15	0.48 a	1.77	1.37	0.56	0.20	73.2	44.2	16.8 a	3.8 a	30.9
Significance	NS	**	NS	NS	NS	NS	NS	NS	***	*	NS
M × C interaction	NS	NS	NS	NS	NS	NS	NS	NS	*	NS	NS
						After Final Harvest					
Treatment			(g/100 g)						(mg/kg)		
Management (M)											
SF w/full N	1.94	0.32	2.22	1.18	0.39	0.20 a	69.4	21.6 b	19.8	5.3	17.6
SH w/full N	2.07	0.33	2.38	1.18	0.34	0.19 b	72.5	19.6 b	19.4	5.4	18.6
SH w/reduced N	2.00	0.30	2.17	1.22	0.36	0.19 b	70.9	29.8 a	18.1	5.6	17.9
Significance	NS	NS	NS	NS	NS	*	NS	**	NS	NS	NS
Cultivar (C)											
Strawberry Festival	2.02	0.33 a	2.25	1.23	0.38 a	0.20 a	75.0 a	24.8 a	18.8	5.3	19.7 a
Camino Real	1.99	0.31 b	2.26	1.16	0.34 b	0.19 b	66.8 b	22.6 b	19.3	5.5	16.4 b
Significance	NS	*	NS	NS	**	*	*	*	NS	NS	**
M × C interaction	NS	**	NS	NS	NS	*	NS	NS	NS	NS	*

SF: summer fallow; SH: sunn hemp; DAT: days after transplanting. The means within the same column followed by the same letter do not differ significantly according to Fisher's least significant difference test at $p \leq 0.05$. NS, *, **, and ***: nonsignificant or significant at $p \leq 0.05$, 0.01, or 0.001, respectively.

Table 4. Strawberry plant above- and below-ground dry biomass after the final harvest and the estimated above-ground plant accumulation of N, P, and K in the organic strawberry field trial in Citra, FL, USA.

Treatment	Above-Ground Biomass (g/plant)	Below-Ground Biomass (g/plant)	N (g/plant)	P (g/plant)	K (g/plant)
Management (M)					
SF w/full N	56.06 a	26.53	1.08 ab	0.18 a	1.25 a
SH w/full N	57.21 a	26.14	1.18 a	0.19 a	1.36 a
SH w/reduced N	48.59 b	24.46	0.97 b	0.15 b	1.05 b
Significance	*	NS	**	*	*
Cultivar (C)					
Strawberry Festival	51.79	25.28	1.04	0.17	1.17
Camino Real	56.12	26.14	1.11	0.17	1.27
Significance	NS	NS	NS	NS	NS
M × C interaction	NS	NS	NS	NS	NS

SF: summer fallow; SH: sunn hemp. The means within the same column followed by the same letter do not differ significantly according to Fisher's least significant difference test at $p \leq 0.05$. NS, *, and **: nonsignificant or significant at $p \leq 0.05$ or 0.01, respectively.

The strawberry leaf tissue nutrient analysis results indicated a limited nutrient contribution from the sunn hemp residues to the following strawberry crop over a long production period of over five months (Table 4). Although sunn hemp has been shown as a potential cover crop in the southeastern U.S to provide N for promoting corn yield in the following season [28], our findings indicated that the role of sunn hemp in increasing soil

N might be restricted by the warm, humid conditions at the time of sunn hemp termination and during the early season of strawberry production. Moreover, flail mowing sunn hemp might also have accelerated the decomposition of the plant residues, resulting in N losses prior to acquisition by the strawberry plants.

3.2. Strawberry Plant Growth

The reduced pre-plant N fertilization treatment resulted in fewer leaves, smaller crowns of the strawberry plants, and a lower leaf chlorophyll content index than the full pre-plant N fertilization treatments at 31 DAT (Table 5). The canopy size was also smaller in the sunn hemp plot with the reduced N fertilization in contrast to the one with the full N fertigation at 31 DAT, but it did not differ significantly from the control without sunn hemp (Table 5). The leaf number and crown diameter remained significantly lower in the reduced fertilization treatment than in the full rate sunn hemp treatment at 51 DAT (Table 5). 'Strawberry Festival' consistently had significantly more leaves than 'Camino Real' at each sampling date from 31 to 100 DAT. A larger canopy size at 65 and 100 DAT and a greater crown diameter at 100 DAT were also observed in 'Strawberry Festival', whereas the leaf chlorophyll content index was significantly lower in 'Strawberry Festival' at 51 DAT (Table 5). As expected, 'Camino Real' had fewer plants with open flowers than 'Strawberry Festival' in the early season flower count. Moreover, 'Strawberry Festival' grown in the reduced pre-plant fertilization treatment exhibited significantly fewer plants with open flowers than the full rate sunn hemp treatment and the summer fallow control. The above-ground biomass assessment after the final harvest revealed a significant reduction in the reduced pre-plant fertilization treatment compared to the full rates with or without sunn hemp, while no difference was detected between the two cultivars, nor in the below-ground biomass (Table 4).

These results of strawberry plant growth parameters indicated that reducing pre-plant fertilization (from 84.0 to 19.8 kg/ha) in the sunn hemp treatment compromised the early growth and development of strawberry plants. In addition, without modifying the fertilization rate for the strawberry crop, growing sunn hemp as a summer rotational cover crop did not show any effects on promoting the growth of strawberry plants, since there were no significant differences between the two full fertilization treatments with and without sunn hemp. Sunn hemp planted in April in the southeastern coastal area of the U.S. and grown over a 90-day period has been reported to produce 8900–13,000 kg/ha of biomass with approximately 135–285 kg/ha of N, at a seeding rate of 13 kg/ha (planted in rows) [13]. In contrast, in the present study, the sunn hemp produced less dry matter (4617 kg/ha) and less N (about 117 kg/ha), with a higher broadcast seeding rate (44.8 kg/ha) in Florida sandy soils but a shorter and later growing period of 67 days from July to September. A relatively high seeding rate was used in our study for weed suppression. A previous report of sunn hemp evaluation in Hawaii showed that the planting date and soil pH exhibited greater impacts on sunn hemp biomass accumulation than the seeding rate [17]. Regardless of the high level of biomass (12,200 kg/ha) and N accumulation (171 kg/ha) in sunn hemp after a 98-day establishment in northern Florida [18], previous studies demonstrated substantial N losses after sunn hemp termination, which resulted in reduced N availability to the subsequent winter corn crop grown in Florida sandy soils.

Table 5. Nutrient management and cultivar effects on leaf number, canopy size, crown diameter, and leaf chlorophyll content index of strawberry plants in the early season (14 November and 4 December 2014, 31 and 51 DAT, respectively), early and peak harvest (18 December 2014 and 22 January 2015, 65 and 100 DAT, respectively), and late season (12 March and 22 April 2015, 149 and 190 DAT, respectively) in Citra, FL, USA.

	Early Season and Peak Harvest														
	Leaf Number Per Plant				Canopy Size (cm)				Crown Diameter (mm)				Leaf Chlorophyll Content Index (SPAD Value)		
Treatment	31 DAT	51 DAT	65 DAT	100 DAT	31 DAT	51 DAT	65 DAT	100 DAT	31 DAT	51 DAT	65 DAT	100 DAT	31 DAT	51 DAT	65 DAT
Management (M)															
SF w/full N	3.4 a	5.3 ab	6.3	11.4 a	13.9 ab	21.7	24.1	26.5	11.9 a	17.0 b	19.3 ab	27.9	45.0 a	49.4	50.8
SH w/full N	3.4 a	5.5 a	6.2	11.6 a	14.9 a	21.8	24.4	27.0	11.8 a	17.9 a	20.0 a	30.9	45.1 a	49.4	50.5
SH w/reduced N	3.0 b	4.7 b	5.7	9.8 b	12.9 b	19.9	23.1	25.5	10.6 b	16.3 b	18.2 b	26.0	41.4 b	47.9	50.7
Significance	*	*	NS	*	*	NS	NS	NS	*	*	*	NS	*	NS	NS
Cultivar (C)															
Strawberry Festival	3.5 a	6.0 a	7.1 a	12.7 a	14.2	21.6	24.9 a	27.1 a	11.7	17.3	19.5	31.3 a	43.0	47.8 b	50.1
Camino Real	3.1 b	4.4 b	5.0 b	9.2 b	13.6	20.7	22.8 b	25.5 b	11.1	16.8	18.9	25.2 b	44.6	50.0 a	51.2
Significance	***	***	***	***	NS	NS	***	**	NS	NS	NS	**	NS	**	NS
M × C interaction	NS	NS	NS	NS	NS	NS	NS	NS	NS	NS	NS	NS	NS	NS	NS

	Late Season						
	Leaf Number Per Plant		Canopy Size (cm)		Crown Diameter (mm)		Leaf Chlorophyll Content Index (SPAD Value)
Treatment	149 DAT	190 DAT	149 DAT	190 DAT	149 DAT	190 DAT	190 DAT
Management (M)							
SF w/full N	15.6	17.9	25.9	28.5	36.3	55.1	41.1
SH w/full N	16.9	16.5	26.7	28.9	42.1	62.6	40.7
SH w/reduced N	14.0	16.7	25.1	29.5	37.9	58.5	41.6
Significance	NS	NS	NS	NS	NS	NS	NS
Cultivar (C)							
Strawberry Festival	15.8	16.8	25.8	29.3	40.0	62.3	42.1
Camino Real	15.1	17.3	25.9	28.6	37.6	55.2	40.1
Significance	NS	NS	NS	NS	NS	NS	NS
M × C interaction	NS	NS	NS	NS	NS	NS	NS

SF: summer fallow; SH: sunn hemp; DAT: days after transplanting. The means within the same column followed by the same letter do not differ significantly according to Fisher's least significant difference test at $p \leq 0.05$. NS, *, **, and ***: nonsignificant or significant at $p \leq 0.05$, 0.01, or 0.001, respectively.

3.3. Strawberry Fruit Yield

With respect to the whole-season yield components (Table 6), 'Strawberry Festival' had greater marketable (by 69.6%) and total fruit numbers (by 36.6%), as well as a higher marketable fruit yield (by 29.3%) than 'Camino Real', whereas 'Camino Real' produced larger berries, as shown by the higher average marketable fruit weight (by 31.5%). The comparisons of soil and nutrient management practices did not indicate any significant difference between the two full pre-plant N fertilization treatments (with or without sunn hemp), while the reduction in pre-plant N fertilization (with sunn hemp) led to significantly lower total fruit number and yield by about 12.7%, relative to the full pre-plant N fertilization treatments (Table 6). This reduction in crop productivity was in line with the decreases in the strawberry plant above-ground biomass and accumulation of N, P, and K after the final harvest (Table 4). However, the marketable fruit number and yield and the average marketable fruit weight did not differ significantly among the nutrient management practices (Table 6).

Table 6. Nutrient management and cultivar effects on the total and marketable strawberry yield components in the organic strawberry field trial in Citra, FL, USA.

Treatment	Marketable Fruit Number (No./plant)	Total Fruit Number (No./plant)	Marketable Fruit Yield (g/plant)	Total Fruit Yield (g/plant)	Average Marketable Fruit Weight (g/plant)
Management (M)					
SF w/full N	11.0	22.9 a	203.1	357.0 a	19.0
SH w/full N	10.6	22.8 a	194.5	353.3 a	18.9
SH w/reduced N	10.3	20.4 b	179.5	309.9 b	18.3
Significance	NS	**	NS	*	NS
Cultivar (C)					
Strawberry Festival	13.4 a	25.4 a	216.9 a	351.8	16.2 b
Camino Real	7.9 b	18.6 b	167.8 b	328.4	21.3 a
Significance	***	***	***	NS	***
M × C interaction	NS	NS	NS	NS	NS

SF: summer fallow; SH: sunn hemp. The means within the same column followed by the same letter do not differ significantly according to Fisher's least significant difference test at $p \leq 0.05$. NS, *, **, and ***: nonsignificant or significant at $p \leq 0.05, 0.01,$ or 0.001, respectively.

In terms of the monthly yield components from December 2014 to April 2015, both sunn hemp treatments (with full or reduced pre-plant N fertilization) showed significantly lower marketable and total fruit numbers and yields in December compared to the summer fallow control (Table 7). Reducing the pre-plant N fertilization (with sunn hemp) also led to decreases in the total fruit number during March–April (by 14.7%) and in the total fruit yield in February (by 12.8%), in comparison to the full pre-plant N fertilization treatments. In addition, both sunn hemp treatments showed lower total fruit yields (by 13.0%) in January than the weedy fallow control. While the yield of 'Strawberry Festival' peaked in February and the fruit size decreased in the late season, 'Camino Real' maintained its fruit size in the late season and produced a greater yield during March–April. Overall, 'Strawberry Festival' outperformed 'Camino Real' in the monthly fruit number and yield from December to February (Table 7).

Table 7. Nutrient management and cultivar effects on monthly marketable and total fruit numbers and yields per plant in the organic strawberry field trial in Citra, FL, USA.

Treatment	Marketable Fruit Number (No./plant)				Total Fruit Number (No./plant)				Marketable Fruit Yield (g/plant)				Total Fruit Yield (g/plant)			
	Dec	Jan	Feb	Mar–Apr	Dec	Jan	Feb	Mar–Apr	Dec	Jan	Feb	Mar–Apr	Dec	Jan	Feb	Mar–Apr
Management (M)																
SF w/full N	0.9 a	2.6	3.2	4.4	0.9 a	4.4	5.2	12.3 a	13.5 a	48.1	67.1	74.3	14.1 a	74.9 a	100.7 a	167.4
SH w/full N	0.7 b	2.3	3.2	4.4	0.7 b	4.1	5.2	12.8 a	9.9 b	42.1	65.5	77.0	10.5 b	64.7 b	99.6 a	178.4
SH w/reduced N	0.6 b	2.5	2.9	4.3	0.6 b	4.4	4.7	10.7 b	8.1 b	41.4	57.1	72.8	8.8 b	65.6 b	87.3 b	148.1
Significance	*	NS	NS	NS	**	NS	NS	*	*	NS	NS	NS	*	*	*	NS
Cultivar (C)																
Strawberry Festival	1.1 a	3.0 a	4.5 a	4.7	1.2 a	4.9 a	6.8 a	12.6 a	15.1 a	47.7 a	83.2 a	70.7	16.2 a	68.9	115.9 a	150.7 b
Camino Real	0.3 b	1.9 b	1.7 b	4.0	0.3 b	3.7 b	3.4 b	11.2 b	5.9 b	40.0 b	43.2 b	78.7	6.2 b	67.9	75.8 b	178.5 a
Significance	***	***	***	NS	***	***	***	*	***	*	***	NS	***	NS	***	*
M × C interaction	NS	NS	NS	NS	NS	NS	NS	NS	NS	NS	NS	NS	NS	NS	NS	NS

SF: summer fallow; SH: sunn hemp. The means within the same column followed by the same letter do not differ significantly according to Fisher's least significant difference test at $p \leq 0.05$. NS, *, **, and ***: nonsignificant or significant at $p \leq 0.05$, 0.01, or 0.001, respectively.

The results of the strawberry yield indicated that the cultivar differences outweighed the effects of soil and nutrient management in both the whole-season and monthly yields of the organically grown strawberries in the present study. The higher early yield observed in the summer fallow control plots compared to the sunn hemp plots might be associated with the tillage effects on the off-season weed control that could have impacted the weed pressure during the following strawberry season. In this study, the summer fallow control plots were tilled twice for weed management before sunn hemp termination. Miler et al. (2014) [29] reported that a summer fallow with roto-tilling twice before bell pepper production resulted in greater nutsedge (*Cyperus* spp.) suppression compared to the no tilling control. Although the whole-season marketable yield was not affected by the sunn hemp treatments, the reduced early yield during the production season in the sunn hemp plots, especially with the reduced pre-plant N fertilization, indicated that relying on the N release from sunn hemp to supplement pre-plant fertilization might not be feasible for promoting plant establishment and early fruiting. It was likely that the fast release of mineralized N from the sunn hemp residues and the potential losses of N from the rootzone either before or after strawberry planting compromised the strawberry plant growth and flowering in the early season, which was particularly evident for 'Strawberry Festival', which flowered earlier than 'Camino Real'. Interestingly, previous studies on tomato (*Solanum lycopersicum* L.) and pepper (*Capsicum annuum* L.) production did not show benefits of sunn hemp residues in fruit yield improvement at a whole-season N application rate of 200 kg/ha until the third year, in comparison with the conventional summer fallow [30]. Hence, the impact of sunn hemp on organic strawberry production deserves to be systematically examined in long-term studies.

Pre-plant N fertilization has been reported to benefit crop establishment and yield performance for soybean (*Glycine max* (L.) Merr.) and corn (*Zea mays* L.) [31–33]; however, the effects of pre-plant N application were inconsistent in several previous studies on Florida strawberries. No difference in either the monthly or total strawberry yield between pre-plant N fertilization at a rate of 56 kg/ha and the control without pre-plant N application was observed [34]. Soilless culture of strawberry plants in coconut coir and pine bark also showed similar early and total marketable fruit yields among different pre-plant N fertilization levels [35]. In contrast, Santos and Ramirez-Sanchez (2009) [36] and Santos (2010) [37] found that pre-plant N fertilization at 56 kg/ha together with S application at 30–64 kg/ha might help increase both the early and total marketable strawberry yields in Florida.

Overall, the strawberry marketable yield in the current study was relatively low because of the high cull percentage compared to previous studies by others [38,39]. Botrytis, anthracnose, and pest damage were observed in this field trial, which were the main causes of the unmarketable yield. Moreover, the lower yield might be related to consecutive frost events encountered in the production season during 19–20 November and 10–16 December 2014, and 28–29 January and 19–21 February 2015 [25], which resulted in a marketable yield reduction due to the extended frost protection with row covers.

4. Conclusions

In this study, growing sunn hemp as a summer rotational crop did not show any growth or yield improvement effects on the winter production of 'Strawberry Festival' and Camino Real' strawberries in Florida sandy soils. Decreasing the pre-plant N fertilization rate by taking into consideration the nutrients, particularly N, provided by the soil incorporation of sunn hemp residues reduced the early and total fruit yields, suggesting that the nutrient contribution by sunn hemp to the strawberry yield performance might be limited. The influence of nutrient management did not vary with the strawberry cultivars used, but overall, 'Strawberry Festival' was shown to be a better yielding cultivar than 'Camino Real' under organic production in this study. Given the environmental impacts of cover crop residue decomposition and nutrient release, long-term studies are needed to better understand the soil nutrient dynamics and strawberry plant nutrient uptake as affected by sunn hemp incorporation. Sunn hemp cropping systems and termination methods and timing also need to be considered to help match the cover crop nutrient release with the subsequent strawberry crop demand for improved crop establishment and fruit yield development.

Author Contributions: Conceptualization, X.Z.; methodology, X.Z., Y.X. and Z.E.B.; formal analysis, Y.X. and X.Z.; investigation, Y.X., Z.E.B. and X.Z.; data curation, Y.X.; writing—original draft preparation, Y.X. and X.Z.; writing—review and editing, Y.X., N.X., Z.E.B., J.K.B., D.M.H. and X.Z.; funding acquisition, X.Z. All authors have read and agreed to the published version of the manuscript.

Funding: This research was funded in part by a National Strawberry Sustainability Initiative grant from the Walmart Foundation that was administered by the University of Arkansas System Division of Agriculture Center for Agricultural and Rural Sustainability.

Data Availability Statement: Data are contained within the article.

Acknowledgments: We thank Buck Nelson and his crew for their assistance with the field work conducted at the University of Florida Plant Science Research and Education Unit.

Conflicts of Interest: The authors declare no conflict of interest.

References

1. U.S. Department of Agriculture. 2021 Certified Organic Survey. Available online: https://www.nass.usda.gov/Surveys/Guide_to_NASS_Surveys/Organic_Production/ (accessed on 25 August 2023).
2. Muramoto, J.; Smith, R.F.; Shennan, C.; Klonsky, K.M.; Leap, J.; Ruiz, M.S.; Gliessman, S.R. Nitrogen contribution of legume/cereal mixed cover crops and organic fertilizers to an organic broccoli crop. *HortScience* **2011**, *46*, 1154–1162. [CrossRef]
3. Parr, M.; Grossman, J.M.; Reberg-Horton, S.C.; Brinton, C.; Crozier, C. Roller-crimper termination for legume cover crops in North Carolina: Impacts on nutrient availability to a succeeding corn crop. *Commun. Soil Sci. Plant Anal.* **2014**, *45*, 1106–1119. [CrossRef]
4. Ebelhar, S.A.; Frye, W.W.; Blevins, R.L. Nitrogen from legume cover crops for no-tillage corn. *Agron. J.* **1984**, *76*, 51–55. [CrossRef]
5. Ranells, N.N.; Wagger, M.G. Nitrogen release from grass and legume cover crop monocultures and bicultures. *Agron. J.* **1996**, *88*, 777–882. [CrossRef]
6. Ozores-Hampton, M. Developing a vegetable fertility program using organic amendments and inorganic fertilizers. *HortTechnology* **2012**, *22*, 743–750. [CrossRef]
7. U.S. Department of Agriculture Natural Resource Conservation Service. Available online: https://www.nrcs.usda.gov/Internet/FSE_DOCUMENTS/nrcseprd331820.pdf (accessed on 19 January 2022).
8. Garland, B.C.; Schroeder-Moreno, M.S.; Fernandez, G.E.; Creamer, N.G. Influence of summer cover crops and mycorrhizal fungi on strawberry production in the southeastern United States. *HortScience* **2011**, *46*, 985–991. [CrossRef]
9. Portz, D.N.; Nonnecke, G.R. Rotation with cover crops suppresses weeds and increases plant density and yield of strawberry. *HortScience* **2011**, *46*, 1363–1366. [CrossRef]
10. Desaeger, J. Meloidogyne Hapla, the northern root-knot nematode, in Florida strawberries and associated double-cropped vegetables. *UF/IFAS Ext.* **2019**, ENY-070. Available online: https://edis.ifas.ufl.edu/publication/IN1224 (accessed on 31 October 2023). [CrossRef]
11. Li, Y.; Hanlon, E.A.; Klassen, W.; Wang, Q.; Olczyk, T.; Ezenwa, I.V. Cover crop benefits for South Florida commercial vegetable producers. *UF/IFAS Ext.* **2022**, SL-242. Available online: https://edis.ifas.ufl.edu/publication/SS461 (accessed on 31 October 2023). [CrossRef]
12. Wang, Q.; Li, Y.; Klassen, W.; Hanlon, E.A. Sunn hemp-a promising cover crop in Florida. *UF/IFAS Ext.* **2022**, SL-306. Available online: https://edis.ifas.ufl.edu/publication/TR003 (accessed on 31 October 2023). [CrossRef]
13. Schomberg, H.H.; Martini, N.L.; Diaz-Perez, J.C.; Phatak, S.C.; Balkcom, K.S.; Bhardwaj, H.L. Potential for using sunn hemp as a source of biomass and nitrogen for the Piedmont and Coastal Plain regions of the southeastern USA. *Agron. J.* **2007**, *99*, 1448–1457. [CrossRef]
14. Bhadha, J.H.; Xu, N.; Rabbany, A.; Amgain, N.R.; Capasso, J.; Korus, K.; Swanson, S. On-farm soil health assessment of cover-cropping in Florida. *Sustain. Agric. Res.* **2021**, *10*, 17–32. [CrossRef]
15. Gaskin, J.; Cabrera, M.; Kissel, D. Predicting nitrogen release from cover crops: The cover crop nitrogen availability calculator. *UGA Coop. Ext.* **2016**. Available online: https://extension.uga.edu/publications/detail.html?number=B1466 (accessed on 31 October 2023).
16. Lynch, M.J.; Mulvaney, M.J.; Hodges, S.C.; Thompson, T.L.; Thomason, W.E. Decomposition, nitrogen and carbon mineralization from food and cover crop residues in the central plateau of Haiti. *Springerplus* **2016**, *5*, 973. [CrossRef] [PubMed]
17. Wang, K.H.; Sipes, B.S.; Hooks, C.R.R.; Leary, J. Improving the status of sunn hemp as a cover crop for soil health and pests management. *Hanai'Ai Newsl.* 2011. Available online: https://projects.sare.org/wp-content/uploads/946541V8-Wang-sunnhemp.pdf (accessed on 30 October 2023).

18. Cherr, C.M.; Scholberg, J.M.S.; McSorley, R. Green manure as nitrogen source for sweet corn in a warm–temperate environment. *Agron. J.* **2006**, *98*, 1173–1180. [CrossRef]
19. Stallings, A.M.; Balkcom, K.S.; Wood, C.W.; Guertal, E.A.; Weaver, D.B. Nitrogen mineralization from 'AU Golden' sunn hemp residue. *J. Plant Nutr.* **2017**, *40*, 50–62. [CrossRef]
20. Agehara, S.; Warncke, D.D. Soil moisture and temperature effects on nitrogen release from organic nitrogen sources. *Soil Sci. Soc. Am. J.* **2005**, *69*, 1844–1855. [CrossRef]
21. Cabrera, M.L.; Kissel, D.E.; Vigil, M.F. Nitrogen mineralization from organic residues. *J. Environ. Qual.* **2005**, *34*, 75–79. [CrossRef]
22. Fan, X.H.; Li, Y.C. Nitrogen release from slow-release fertilizers as affected by soil type and temperature. *Soil Sci. Soc. Am. J.* **2010**, *74*, 1635–1641. [CrossRef]
23. O'Connell, S.; Shi, W.; Grossman, J.M.; Hoyt, G.D.; Fager, K.L.; Creamer, N.G. Short-term nitrogen mineralization from warm-season cover crops in organic farming systems. *Plant Soil* **2015**, *396*, 353–367. [CrossRef]
24. Quemada, M.; Cabrera, M.L. Temperature and moisture effects on C and N mineralization from surface applied clover residue. *Plant Soil* **1997**, *189*, 127–137. [CrossRef]
25. Florida Automated Weather Network. Available online: http://fawn.ifas.ufl.edu (accessed on 19 January 2022).
26. Li, J.; Zhao, X.; Maltais-Landry, G.; Paudel, B.R. Dynamics of Soil Nitrogen Availability Following Sunn Hemp Residue Incorporation in Organic Strawberry Production Systems. *HortScience* **2021**, *56*, 138–146. [CrossRef]
27. Agehara, S.; Hochmuth, G. Fertilization of strawberries in Florida. *UF/IFAS Ext.* **2023**, CIR-1141. Available online: https://edis.ifas.ufl.edu/publication/CV003 (accessed on 31 October 2023). [CrossRef]
28. Balkcom, K.S.; Reeves, D.W. Sunn-hemp utilized as a legume cover crop for corn production. *Agron. J.* **2005**, *97*, 26–31. [CrossRef]
29. Miller, M.R.; Dittmar, P.J.; Vallad, G.E.; Ferrell, J.A. Nutsedge (*Cyperus* spp.) control in bell pepper (*Capsicum annuum*) using fallow-period weed management and fumigation for two years. *Weed Technol.* **2014**, *28*, 653–659. [CrossRef]
30. Avila, L.; Scholberg, J.; Roe, N.; Cherr, C. Can sunn hemp decrease nitrogen fertilizer requirements of vegetable crops in the Southeastern United States? *HortScience* **2006**, *41*, 1005. [CrossRef]
31. Osborne, S.L.; Riedell, W.E. Starter nitrogen fertilizer impact on soybean yield and quality in the northern Great Plains. *Agron. J.* **2006**, *98*, 1569–1574. [CrossRef]
32. Touchton, J.T.; Rickerl, D.H. Soybean growth and yield responses to starter fertilizers. *Soil Sci. Soc. Am. J.* **1986**, *50*, 234–237. [CrossRef]
33. Vetsch, J.A.; Randall, G.W. Corn production as affected by tillage system and starter fertilizer. *Agron. J.* **2002**, *94*, 532–540. [CrossRef]
34. Agehara, S.; Santos, B.M.; Whidden, A.J. Nitrogen fertilization of strawberry cultivars: Is preplant starter fertilizer needed? *Univ. Fl. IFAS Ext.* **2007**, HS1116. [CrossRef]
35. Cantliffe, D.J.; Castellanos, J.Z.; Paranjpe, A.V. Yield and quality of greenhouse-grown strawberries as affected by nitrogen level in coco coir and pine bark media. *Proc. Fla. State Hortic. Soc.* **2007**, *120*, 157–161.
36. Santos, B.M.; Ramirez-Sanchez, M. Effects of preplant nitrogen fertilizer sources on strawberry. *Proc. Fla. State Hortic. Soc.* **2009**, *122*, 240–242. [CrossRef]
37. Santos, B.M. Effects of preplant nitrogen and sulfur fertilizer sources on strawberry. *HortTechnology* **2010**, *20*, 193–196. [CrossRef]
38. Fernandez, G.E.; Butler, L.M.; Louws, F.J. Strawberry growth and development in an annual plasticulture system. *HortScience* **2001**, *36*, 1219–1223. [CrossRef]
39. Hargreaves, J.C.; Adl, M.; Warman, P.R.; Rupasinghe, H.P. The effects of organic and conventional nutrient amendments on strawberry cultivation: Fruit yield and quality. *J. Sci. Food Agric.* **2008**, *88*, 2669–2675. [CrossRef]

MDPI AG
Grosspeteranlage 5
4052 Basel
Switzerland
Tel.: +41 61 683 77 34

Horticulturae Editorial Office
E-mail: horticulturae@mdpi.com
www.mdpi.com/journal/horticulturae